LOUDSPEAKER RECIPES

BOOK 1:
FOUR TWO-WAY SYSTEMS

LOUDSPEAKER RECIPES

BOOK 1:

FOUR TWO-WAY SYSTEMS

By
Vance Dickason

Audio Amateur Press
Publishers
Peterborough, New Hampshire

FIRST EDITION
1994

Distribution Agents:
Old Colony Sound Lab
Post Office Box 243
Peterborough, New Hampshire 03458-0243 USA

Library of Congress Card Catalog Number 94-071764

ISBN: 1-882580-04-4

Dedication

This work is dedicated to the memory of my father, Raymond D. Dickason, who, prior to his death in 1978, had two goals in mind for the rest of his life. The first and most important was to pursue an idea for producing a low-energy fusion reaction—cold fusion—which would supply a source of cheap and unlimited power for the world; and failing that, to join the Peace Corps. This dedication is for him and for all of us who feel compelled to find our heroic purpose in life.

Introduction

The most frequently asked question since *The Loudspeaker Design Cookbook* (*LDC*) was published in 1975 is, "Where are the recipes?" Although the *LDC* was originally intended to be a précis of loudspeaker design technology and not an applications manual, it has always been my goal to write a book which demonstrated the use of all this information in a detailed portrayal of realized designs.

However, rather than illustrate the design process using the manual and hand calculator methods described in the *LDC*, my idea was not only to make this book a series of construction articles, but also to investigate the mysteries of loudspeaker quality with the latest in computer-aided engineering (CAE) software. Not only does CAE software significantly speed up the design chore (an absolute necessity when simultaneously designing a large number of loudspeakers), but it gives us invaluable insight into what happens in the design process as well. You can readily apply this information even without such sophisticated and sometimes very expensive tools.

My task, however, turned out to be formidable and time-consuming. I began planning the work over seven years ago, at about the time Ed Dell founded (and I began editing) *Voice Coil*, a monthly newsletter for loudspeaker manufacturers. Originally, I intended to present twelve loudspeaker designs in a single volume, each representing a different design format used in the industry. I hoped to include two-, three-, and four-way speakers, as well as a musical instrument type as a keyboard monitor to accompany my Roland synthesizer.

As time progressed, I chose the initial driver complements, asked various driver manufacturers for participation, and collected, over the course of a full year, the various transducer samples. Testing the amazingly wide variety of drivers led to my final choices, and at last I was able to explore the box designs through the use of a new (new four years ago, that is) and highly accurate computer box design program, LEAP (Loudspeaker Enclosure Analysis Program) 3.1. During another full year, I completed the design formats and box designs.

Thus began the construction phase requiring cutting and assembling 30 different cabinets. Once these raw MDF shells were complete, the nearly equal task of box finishing began, again for 30 enclosures. Since this work could not be a full-time occupation, it took another three years to finish the cabinets and get them ready for final crossover design.

All during this period, many changes were taking place in the fields of computer-aided loudspeaker design software and computer-based acoustic measurement equipment. As I reviewed each of the new programs and new program versions in *Voice Coil*, I could see that it was becoming more practical to do the entire loudspeaker design, box dimensions, and crossover design on the computer and then determine if the design performed exactly as described by the software when it was finally prototyped and measured.

At the same time, I was personally involved in the development of one of the finest examples of professional loudspeaker development software available in this industry, Audio Teknology's LEAP 4.5. While I was working for ATI, company president and founder Chris Strahm not only encouraged me to offer suggestions and influence the direction the software was taking, but also provided me with an invaluable education in the process. (I have since left my position at ATI and am an independent consultant. ATI has been restructured and LEAP is now sold by LinearX Systems, Inc.)

Parallel to the development of computer design tools such as LinearX's LEAP, Peter Schuck's XOPT, David Clark's SPEAK, SpeakEasy's LFD, and Audiosoft's CALSOD, an amazing series of affordable ($1,000–$10,000) measurement tools has appeared which five years ago didn't exist. I have reviewed a number of these analyzers in the pages of *Voice Coil* and developed a familiarity with the attributes and weaknesses of these reasonably low cost but highly effective instruments. The list includes the Ariel SYSid, the DRA MLSSA, the Techron TEF 20, the Audio Precision System 1 DSP, and the LinearX LMS (Loudspeaker Measurement System). My real excitement came when we were able to interface the measurements from this type of equipment with the computer modeling programs. We could load the computer-generated files from any loudspeaker measurement into a CAE program, making truly accurate design modeling possible.

In order for my 12-design project to finally get under way, a number of things had to happen. First I limited its scope. Instead of one large book with twelve designs, I chose to produce three books, each with four designs. This first result is *Loudspeaker Recipes, Book 1: Four Two-Way Systems*.

The next and most important factor in enabling this work was the release of the latest version of the LEAP CAE software. Version 4.5 (for which I was a beta test station) finally put together the complete set of tools to make my project come together in the way I originally planned it. Although its conceptualization took place in 1987, it wasn't until the fall of 1992 that all the computer measurement and design tools were advanced to the degree that this book could be possible.

The design process I used for these speakers is as follows. Each of the four represents a different set of tradeoffs and techniques for making a workable two-way design. I performed all the design work on the computer and none of the speakers was "tweaked" following the design process. All four designs performed within a very narrow range of tolerance exactly as predicted by the computer design process. If there was any doubt, this book definitely confirms the viability of computer design tools. However, we must not lose sight of the fact that computers do not design loudspeakers—they are only tools which greatly facilitate and speed up the design process. What was once considered nearly a "black art" has now become a much more objective pursuit, but one that still requires the magic of creativity.

The drivers and components in this book's four projects are all off-the-shelf products which are readily obtainable by home constructors. While some extremely fine drivers are available in this way, a majority of the finished speakers sold commercially utilize custom driver designs that represent variations from standard units. Even so, you the amateur can still produce excellent quality loudspeakers with the appropriate application of well thought out design concepts and without access to "tweaked" versions.

If I were doing the same kind of work for a manufacturer in the capacity of a design consultant (which I frequently do), the designs would understandably look somewhat different. These systems were built with readily available parts and with no limits dictated by a target price or retail sales margin. In that respect, they are purely design exercises intended to demonstrate useful principles and techniques.

All the designs in this book are protected by copyright. Anyone is welcome to reproduce these designs for personal use. However, production of these designs for sale or any other commercial use, without the author's written permission, will be regarded as copyright infringement and treated accordingly. Anyone wishing to duplicate these designs commercially is invited to contact me regarding licensing arrangements and/or custom versions of these products.

Design Philosophy

I followed a number of design principles throughout this book and applied them to each of the four speaker designs. These principles, when appropriately applied, yield what is usually regarded as a high quality loudspeaker. They are:

1. Loudspeakers should have a flat frequency response (both the on-axis response and the power response).

A flat response does not necessarily make a great loudspeaker, but few speakers with widely varying response (±5dB or more) are judged to be superior. In other words, the race doesn't necessarily go to the swift, but that's the way to bet. This opinion has been expressed by many audio engineers and members of the audio review press.

Dr. Floyd Toole wrote a series of excellent papers for the *JAES* (*Journal* of the Audio Engineering Society) during his employment at the NRC (the Canadian National Research Council; Dr. Toole is now Vice President of Acoustic Research at JBL) which successfully attempted to correlate listener preference in scientifically designed listening tests with the spectral content of a loudspeaker's frequency response.

His group's conclusions, as Dr. Toole stated, were much like an affirmation of motherhood. "Given the proper circumstances, experienced listeners with normal hearing prefer loudspeakers with wide bandwidth, flat and smooth amplitude response, and uniformly wide dispersion".[1] Daniel Queen, in a 1979 *JAES* paper,[2] also concluded that wide-dispersion speakers were desirable providing they maintained spectral uniformity (were flat).

Martin Colloms noted in his book *High Performance Loudspeakers*[3] that "successful models have been designed virtually for free-field use and are best auditioned on open stands as far away as is practical from room walls or corners. Such systems appear most natural if their anechoically measured axial response is essentially uniform with frequency, provided that certain other requirements are met, namely that the system should employ direct radiator drivers possessing a good dispersion over the forward ±30° angle, and that the levels of distortion and delayed resonance coloration should be low." Colloms further suggests a standard of deviation for high quality speakers to be flat within ±2dB with respect to the on-axis curve over ±30° laterally, and ±10° vertically.

The review press has also stated a preference for flat, wide-dispersing speakers. Don Keele (one of the few engineer/reviewers), in a recent issue of *Audio* magazine,[4] maintained that ". . . the on-axis response should be smooth, because it defines the spectral balance of the sound that first arrives at the listener and so is of greatest subjective importance in judging timbre. Smooth power response is important because the room's sound and associated reverberation depend heavily on the off-axis response of the speaker and its power response. The total sound of a speaker depends not only on the direct sound reaching the listener but also on the later reflected sound.

"Assuming a properly designed listening room, a speaker with smooth on-axis response will sound best. Accurate stereo image depends on absolute symmetry in your listening setup, on close right/left matching of the loudspeakers, and very uniform on- and off-axis horizontal response. Low distortion, particularly at high levels and low frequencies, is a requirement for realistic wide-range sound reproduction."

John Atkinson, editor of *Stereophile* magazine, did a very interesting study[5] of loudspeaker quality and spectral flatness. This short dissertation described some 74 loudspeaker models reviewed in *Stereophile* from 1990-1992. Atkinson divided the 74 speakers into five categories depending on their weighted deviation from a flat frequency response as follows:

Group	Deviation	# Spkrs	# Recommended	% Recommended
1	<2dB	17	14	82.4
2	2–2.5dB	13	10	76.9
3	2.5-3	17	13	76.5
4	3–3.7	12	8	66.6
5	>3.7	15	5	33.3

From these rankings it's apparent that although there were recommended speakers in all classes ("A" recommended speakers appeared only in groups 1 and 3), the manufacturer had a much better chance of making the grade if deviations were kept to less than ±3dB.

His conclusions were essentially that "flatness of midrange and treble response is a good thing to have in a loudspeaker, (but) it doesn't in itself mean that they will sound good, or lead to an automatic recommendation by *Stereophile*....The best loudspeakers, in my opinion, combine a flat on-axis midrange and treble with an absence of resonant colorations, a well controlled high-frequency dispersion, excellent imag-

References

1. Toole, Floyd E., "Loudspeaker Measurements and Their Relationship to Listener Preferences: Part 2," JAES, May 1986.
2. Queen, Daniel, "The Effects of Loudspeaker Radiation Patterns on Stereo Imaging and Clarity," JAES, May 1979.
3. Colloms, Martin, *High Performance Loudspeakers*, Fourth Edition, John Wiley & Sons, 1991, p. 316.
4. Keele, Don B., "Equipment Profile: JBL L7 Speaker," *Audio*, December 1992.
5. Atkinson, John, "JA on Loudspeakers," *Stereophile*, October 1991. Vol. 14, No. 10, pp. 205-212.

ing, precision an optimally tuned bass, and also play loud and clean without obtrusive compression." I couldn't have said it better.

In practical terms, the criteria given above determined two design goals for this book's loudspeakers. The first was to achieve a frequency response deviation no greater than ±2dB on-axis. The second, to maintain as nearly as flat possible a response horizontally at 15° and 30° off-axis. Because all four loudspeakers used 6-7″ woofers and since most 6.5″ woofers typically roll off appreciably off-axis at around 3kHz, this dictated an optimum crossover frequency of approximately 2kHz in order to achieve the latter goal.

2. Loudspeakers should be as free as possible from distortion and compression effects.

To avoid distortion and compression in the woofer section of a two-way speaker we must design the enclosure so the woofer remains in its linear excursion region throughout most of the dynamic operating range. This goal is relatively easy to accomplish using computer modeling and requires the designer to choose a box size that avoids excessive excursion. However, high-frequency drivers can also be an annoying source of distortion. The chosen 2kHz crossover frequency is a potential source for distortion caused by excessive tweeter excursion. 2kHz is a dangerously low crossover frequency for most available 1″ dome tweeters to remain free of over excursion.

Here the problem is closely related to the crossover attenuation slope. While opinions vary in the industry, each favoring the sonic qualities of different classic filter functions, the low crossover frequency criterion demands maximum attenuation at that frequency if distortion is to be kept to a minimum. For this reason, I chose fourth-order acoustic slopes for the crossover designs in all four systems. Although many designers favor a first-order acoustic crossover slope, this type of network is generally only suitable for a few specifically designed tweeters such as the Dynaudio Esotar (used in the Sonus Faber Extrema). First-order crossover slopes are much more appropriate for three-way designs.

3. High quality loudspeakers should present an appropriate sonic image in the stereo soundfield.

Several design criteria are necessary for good image quality in a loudspeaker, the most obvious being close frequency response matching of left and right stereo pairs. This is essential, and a relatively straightforward matter of carefully matching components. Wide dispersion in the mid- and upper-range is also an important factor, and has already been discussed. Another criterion which I believe affects image quality is having closely symmetrical (or even precisely asymmetrical) high-pass and low-pass crossover slopes with both drivers electrically in-phase. To maintain this goal with both high-pass (tweeter) and low-pass (woofer) speakers conventionally mounted on the same baffle board, the design must somehow take into account the physical time delay between the woofer and tweeter. Each of the four designs in this book presents a different design procedure for coping with the driver delay difference presented by conventional mounting techniques. While I will explain these design procedures in more detail later, they are as follows:

- Speaker 1: A sloped baffle is used to reduce inter-driver delay from 120μs to 60μs. Using a fourth-order L-R crossover at 2.2kHz, 60μs delay causes only a 1dB variance from flat.

- Speaker 2: A sloped baffle is used to reduce inter-driver delay from 160μs to 100μs. Using a fourth-order L-R and overlapping the high-pass and low-pass sections (the high-pass at 2.2kHz and the low-pass at 2.4kHz) rendered the response flat to 1dB in the crossover region.

- Speaker 3: An asymmetrical network with a second-order Butterworth low-pass section at 2.6kHz and a fourth-order L-R section at 3kHz produced a flat response within 1dB through the crossover region with 95μs of interdriver delay.

- Speaker 4: The physical configuration of the drivers resulted in a interdriver delay of only 60μs. A standard L-R filter at 2.2kHz rendered the network flat to within 1dB.

Measurement Procedures and Equipment

I used the following test equipment and computer software for the designs in this book:

Acoustic Test Instruments:

1. LinearX LMS gated sinewave analyzer
2. DRA Laboratories MLSSA FFT analyzers

Microphones

1. LinearX LMS microphone (software calibrated to a B&K 4133)
2. ACO Pacific model 7012 microphone and PS9200 power supply/preamp

Computer Design Software

1. LinearX LEAP v. 4.5

Virtually all measurements made in this book were made outdoors using LMS. Full-range anechoic measurements were created by using two separate measurement methods. Low-frequency measurements from 10Hz to approximately 300–400Hz were produced by making ungated groundplane measurements. These low-frequency measurements were then spliced together with free-field gated sinewave SPL measurements made on a five-foot tower. The tower measurements were made from 300Hz–40kHz.

All measurements were made with the microphone placed one meter from the speaker. For the gated 300Hz–40kHz measurements, the one-meter distance is considered minimal for placing the microphone in the far field for speakers of small baffle dimensions.[1] However, placing the microphone at a greater distance (such as two meters) limits the low-frequency response of the measurement to a higher frequency than is desirable, due to the nature of gated measurements.

Since I consider it optimum to splice the groundplane measurements at a frequency no higher than 300–400Hz, this precluded using a distance greater than one meter for the gated tower measurements. However, since three of the four speakers in this book have baffles with their longest dimension less than 17″, the one meter distance was not much of a compromise (the far field is loosely defined as beginning approximately 3× the largest dimension of the baffle). Although it's desirable to be back farther than one meter in a measurement situation, most physical layouts used by a majority of loudspeaker engineers preclude using longer microphone-to-speaker distances. Fortunately, the consequences of shorter measurement distances are usually not dire.

List of Contributors

I received assistance from a number of people in the industry, without whose gracious cooperation this book would have been nearly impossible. "The staff" and contributors were as follows:

CABINET CONSTRUCTION:
Larry Arntson
Jason Dickason

SUBWOOFER AND
STAND CONSTRUCTION:
The Speaker Clinic
1 S.E. 47th
Portland, OR 97215
(503) 230-0366

CONSTRUCTION FACILITY:
Design Sound
12805 S.E. 27th Place
Portland, OR 97222
(503) 786-9493

NOTE: Both The Speaker Clinic and Design Sound can supply prerouted front baffles for any of the speakers described in this book.

Reference
1. Struck and Temme, "Simulated Free Field Measurements," 93rd AES Convention, October 1992, Preprint #3397

DRIVER SUPPLIERS:

ACI (Audio Concepts, Inc.)
Dynaudio (Madisound Speaker Components)
Focal (Orca Design & Manufacturing)
Morel Acoustics USA
Scan-Speak (A & S Speakers)
Vifa

CROSSOVER PARTS:

I.T. InterTechnik USA
PO Box 186
Kent, WA 98035
(206) 852-4049/FAX (206) 852-1203

North Creek Music Systems
PO Box 1120
Old Forge, NY 13420
PHONE/FAX (315) 369-2500

Orca Design & Manufacturing
1531 Lookout Dr.
Agoura, CA 91301
(818) 707-1629/FAX (818) 991-3072

Solen Inc.
4470 Thibault Ave.
St-Hubert, PQ J3Y 7T9 Canada
(514) 656-2759/FAX (514) 443-4949

OTHER COMPONENTS:

Audio Concepts, Inc.
901 S. 4th St.
La Crosse, WI 54601
(608) 784-4570/FAX (608) 784-6367

A & S Speakers
3170 23rd St.
San Francisco, CA 94110
(415) 641-4573/FAX (415) 648-5306

Madisound Speaker Components
PO Box 44283
Madison, WI 53744-4283
(608) 831-3433/FAX (608) 831-3771

Mahogany Sound
2610 Schillingers Rd. #488
Mobile, AL 36695
(205) 633-2054

Marchand Electronics, Inc.
PO Box 473
Webster, NY 14580
(716) 265-4930/FAX (716) 265-1614

Meniscus
2575 28th St. SW #2
Wyoming, MI 49509
(616) 534-9121/FAX (616) 534-7676

Monster Products, Inc.
274 Wattis Way
S. San Francisco, CA 94080
(415) 871-6000/FAX (415) 871-6592

Orca Design & Manufacturing
(see above)

TDM Design
Rt. 1, Box 573
Hillsboro, OR 97124
(503) 647-5957/FAX (503) 647-5953

CONTENTS

LOUDSPEAKER RECIPES

BOOK 1:

FOUR TWO-WAY SYSTEMS

Chapter One

A SIMPLE VENTED TWO-WAY

1.10 Overview

The first of the four designs in this book, based on a 6.5" woofer and a 1" dome tweeter, is, perhaps, the ideal two-way speaker format. Although general engineering consensus holds that an 8" woofer provides the best trade offs for two-way speakers, it has sufficient piston area to provide authoritative bass and cone directivity, thus allowing effective crossover design. A 6.5" woofer can also provide reasonably good bass extension.

Additionally, it has better overall off-axis response, which allows you to either cross the tweeter over at a higher frequency (perhaps lowering distortion) or produce a potentially superior power response. While this is something of a generalization, two-way 6.5" formats, such as the Celestion SL600, have been favorites of audiophiles for many years.

For this design I used the Vifa P17WJ-00-08 woofer and the Scan-Speak D2905 tweeter.

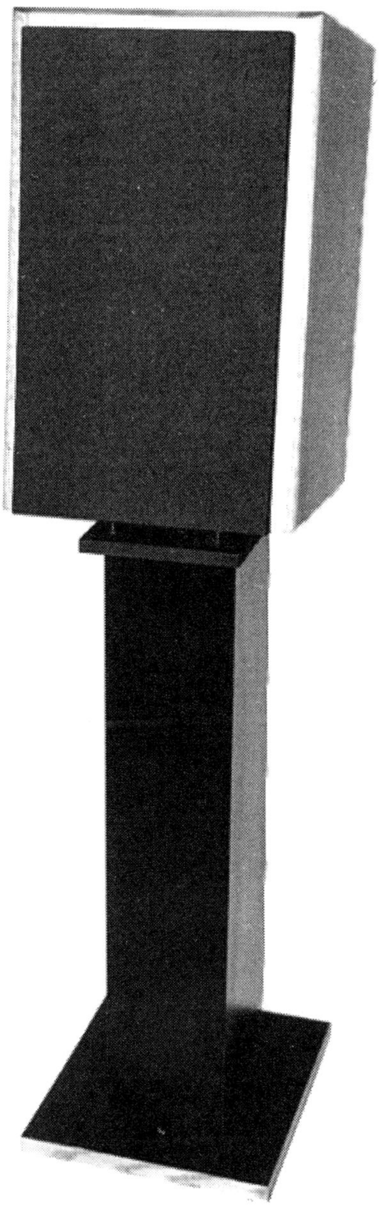

1.20 The Woofer

The Vifa P17WJ-00-08, originally released in February 1988, is a fairly old design in the field of off-the-shelf, original equipment manufacturer (OEM) products. It is in fact, a classic. Well-known manufacturers, such as Thiel Audio, PSB, Definitive Audio, Unity Audio, and Conrad Johnson Synthesis, have used stock OEM and custom versions of the P17WJ.

This design uses a cast magnesium basket, a polypropylene cone and soft-poly dust cap, a rubber surround, and a flat spider. The two-layer voice coil is wound on a 32mm-diameter aluminum former. Vifa rates the P17WJ's sensitivity at 88dB, and gives it an IEC Standards 268-5 nominal power rating of 70W.

For this project, I used four sample woofers. My first step was to break them in by running a low-frequency sine wave (20-25Hz) at a sufficient level to exercise the suspension system for 24 hours. Next I measured the woofers' impedance using LinearX's *Loudspeaker Measurement System* (LMS) analyzer with the drivers mounted in their operating positions in the open air without a baffle (i.e., a standard free-air measurement). The test jig I used rigidly affixes the driver to an nonobstructive wood framework, which prevents the speaker basket from moving during the testing. I made the impedance measurements from 10Hz-40kHz using a high-resolution 550-point sweep.

A second sweep was performed, with the woofer face down in a sealed, 0.25 ft^3 test box to determine compliance and V_{AS}. Next, I measured driver resistance (DCR) using a Fluke 8060 multimeter. By subtracting the test cable's resistance from the measurement, I processed the impedance data further. After shorting the ends of the impedance

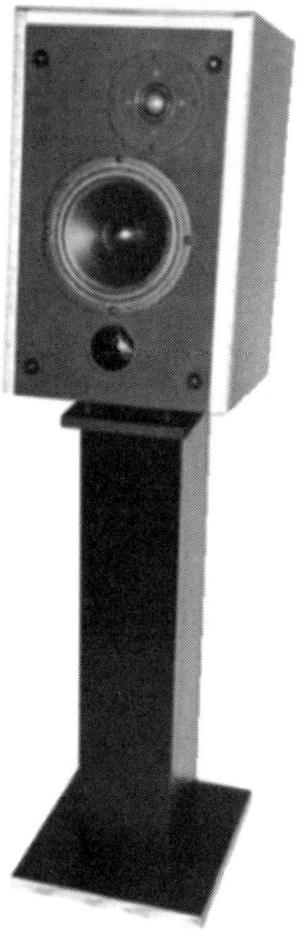

test cable and running an additional sweep, I subtracted the resulting cable-resistance sweep data file from each of the impedance measurements using the A-B command in the LMS software. *Figure 1.1* depicts one sample's free-air impedance.

To measure the drivers' parameters, I transferred the processed impedance curves, plus one free-air and one test-box measurement for each of the four samples, from the LMS into the LEAP 4.5 (Loudspeaker Enclosure Analysis Program) software, also from LinearX (formerly Audio Teknology, Inc.). An option in the export/import process allows you to instruct LEAP to calculate the phase curves for each measurement. No matter which analyzer you use, it's a good practice to take advantage of LEAP's calculated-phase routine.

Although the impedance phase from other sine-wave analyzers (such as the dual-channel Audio Precision System 1) is completely accurate, the various fast fourier transform (FFT) analyzers tend to be only semiaccurate, depending on the resolution and bandwidth of the measurement. LEAP's sophisticated and modified Hilbert transform routine generally provides more accurate data than the measured phase from the truncated impulse of various FFT-type acoustic analyzers, unless the measurement is made under tightly controlled conditions, such as in an anechoic chamber. This applies to both impedance (which suffers not from truncation, but from linear resolution errors) and especially sound pressure level (SPL) measurements.

I used the LEAP Speaker Parameter Measurement utility to create the parameter list in *Table 1.0* for the four P17WJ woofers.

Although the variations might seem significant, they are not. Simulating the woofer's performance using all four drivers' parameter sets in the same size box, (such as the 0.3 ft^3 QB3-type enclosure

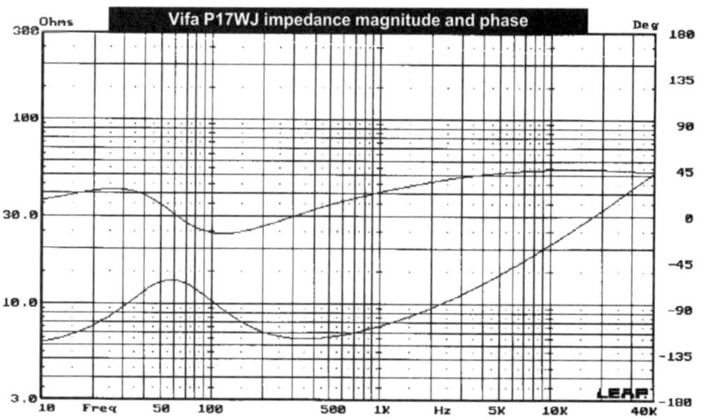

— Free air impedance

FIGURE 1.1

Table 1.0

	A	B	C	D
Z_{NOM}	8	8	8	8
R_E (Ω)	5.61	5.65	5.67	5.59
S_d(M^2)	0.130	0.130	0.130	0.130
Bl (TM)	5.55	5.998	6.069	5.95
V_{AS} (ltr)	19	23	22	23
C_{MS} (μM/N)	792	971	915	971
M_{MS} (gm)	9.02	11.72	12.44	11.72
f_S (Hz)	51.8	47.2	47.2	47.2
Q_{MS}	0.965	0.984	0.898	0.886
Q_{ES}	0.614	0.546	0.568	0.548
Q_{TS}	0.375	0.351	0.348	0.338
SPL(2.83V)	90.79	89.29	88.88	89.32
L_E (1kHz)	0.6	0.59	0.56	0.56
(20kHz)	0.18	0.2	0.2	0.2

illustrated in *Fig. 1.2*) produces nearly overlapping curves for each speaker sample.

The one disadvantage to building your own speaker is the problem of driver consistency. If you purchase only two drivers, it's impossible to use the quality-control selection processes used by many manufacturers. The manufacturers of the drivers used provide consistent product, and the variations between units are usually minor, as you can see from *Table 1.0.*

Vifa designed the P17WJ for use in a vented enclosure. Given their parameters, the LEAP Quick Cabinet utility, (a basic box-calculator) indicated an f_3 of about 100Hz in a sealed box, and 50Hz or so in vented alignments. It also recommended three vented configurations: a quasi-third-order Butterworth (QB3), a sub Chebychev/Bessel, and an alignment developed by LEAP's programmer Chris Strahm, referred to as extended-bass shelf (EBS). The QB3 and Sub Chev/Bes alignments both specified a 0.34 ft.^{3} box tuned to 53Hz and 51Hz, respectively; the EBS called for a 0.53 ft.^{3} box size tuned to 43Hz. Since the QB3 and Sub Chev/Bes were so close in terms of box volume and tuning frequency, I only considered the QB3 and EBS configurations as possibilities.

Figure 1.3 gives the half-space simulation for both vented alignments, plus a higher tuning for the EBS to 50Hz. The QB3 simulation has an f_3 of 59Hz, the lower-tuned EBS (6" vent length, 2"ID) is at 47Hz, the higher-tuned EBS (4" vent length, 2"ID) is at 48Hz. The group-delay curves in *Fig. 1.4*, and the cone-excursion curves in *Fig. 1.5*, suggest the higher-tuned EBS alignment, with fewer excursion demands, is a good compromise between damping and bass extension. The largest vent able to fit into a 0.53 ft.^{3} box is 2" in diameter. A 3-inch diameter vent (ABS tubing is available only in 1" increments) tuned to 43Hz is nearly 15" in length; and 10" when tuned to 50Hz.

To further investigate the EBS-type alignment, I performed two sets of simulations, one at 1W input power and another at 10W. Both used the three different box tunings corresponding to 4", 5", and 6-inch vent lengths for a 2"ID pipe. This kind of simulation gives you a better understanding of the dynamic changes a particular box design produces in low-end response. The graphs produced were:

1W	10W	Graph Type
Fig. 1.6	Fig. 1.13	Freq. Response (groundplane)
Fig. 1.7	Fig. 1.14	Cone Excursion
Fig. 1.8	Fig. 1.15	Group Delay
Fig. 1.9	Fig. 1.16	Impedance
Fig. 1.10	Fig. 1.17	Port Excursion (mm)
Fig. 1.11	Fig. 1.18	Cone Velocity (m/s)
Fig. 1.12	Fig. 1.19	Port Velocity (m/s)

The most important indication of dynamic performance in this curve set is cone excursion. As *Fig. 1.14* shows, with the box tuned to 50Hz the driver excursion is about 4.6mm at 35Hz at 10W. The linear excursion limit (X_{MAX}), for this woofer is 4mm. Assuming acceptable harmonic distortion up to 15% beyond X_{MAX}, this woofer operates in a linear range when the cone travels no farther than 4.6mm forward or backward from its center (rest) position.

Since this excursion distance occurs at a frequency slightly below 40Hz, and since 95% of program material is above this frequency, the P17WJ in this configuration should perform up to the 100dB SPL indicated in *Fig. 1.13* without signs of severe stress and distortion. (The 105-106dB midband level in *Fig. 1.13* minus 6dB for groundplane effect gives the 100dB SPL.)

The rapid increase of cone excursion from 35-100Hz indicates that playing low-frequency material at high volumes could easily bottom out the voice coil to the backplate. The P17WJ's flat spider gives the suspension a progressive character. The return force of the suspension system increases at the extremes of excursion and reduces the propensity to bottom out in practice. Although it is a tedious procedure, you can manually graph the suspension system's exact excursion characteristics, and alter LEAP's analysis to reflect the more progressive spider.

Overall, the P17WJ performs well at reasonably loud listening levels. I consider any single woofer which remains linear at 100dB to be satisfactory for home audio. Unless, of course, your goal is a high-performance speaker with a dynamic range linear to 110dB or so, although the extra 10dB is hard to come by in single OEM drivers for the home market. The remaining figures, which are fairly self-explanatory, will give you a sense for the different characteristics of dynamic change.

Once I'd constructed the prototype cabinet and mounted the drivers, I confirmed the low-frequency simulations by making outdoor groundplane measurement, (*Fig. 1.20*). As with the simulation, I performed this test using 4", 5", and 6-inch vent lengths. Although the frequency scaling is different (20-500Hz rather than 10Hz-1kHz), you can easily see that the LEAP simulations were quite accurate and the measured low-end response is exactly as predicted. (The SPL level for a groundplane measurement is 6dB greater than for an anechoic measurement. See *Chapter 8* of *The Loudspeaker Design Cookbook*, Fourth Edition.[1])

The full-range anechoic measurement of the mounted P17WJ is shown in *Fig. 1.21*. I arrived at this by combining groundplane and gated-anechoic measurements. I first made the groundplane measurement from 10-400Hz, and then scaled the groundplane data to the level of an anechoic

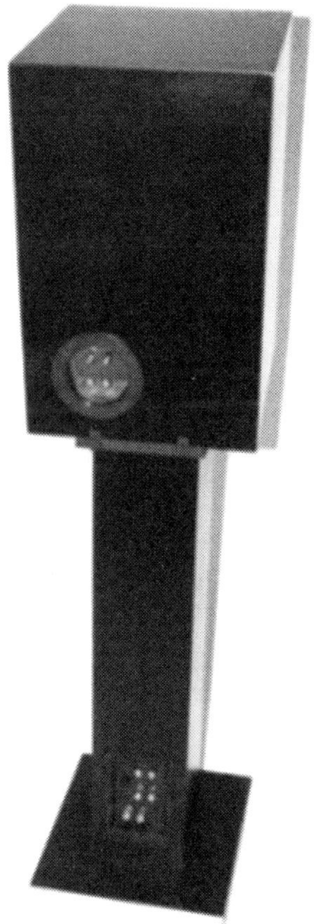

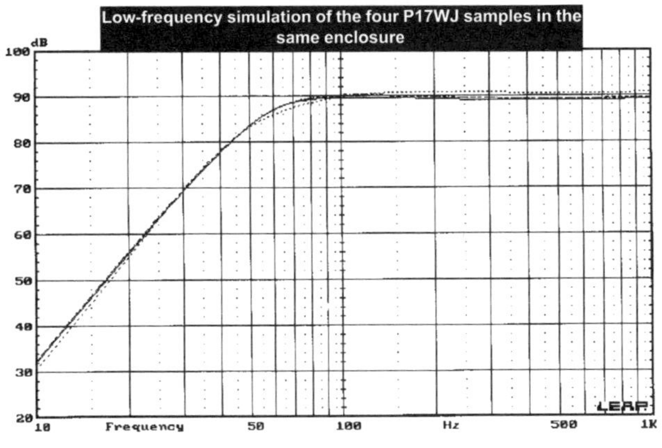

—— Sample A QB3 box ······· Sample B QB3 box - - - Sample C QB3 box -·-· Sample D QB3 box

FIGURE 1.2

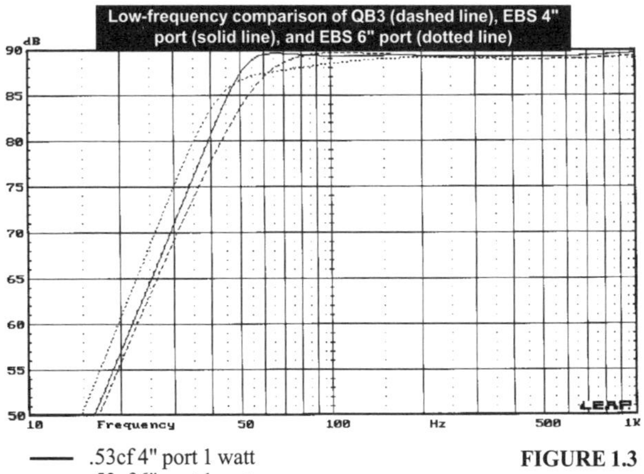

—— .53cf 4" port 1 watt
······· .53cf 6" port 1 watt
- - - .34cf (QB3) 6" port 1 watt

FIGURE 1.3

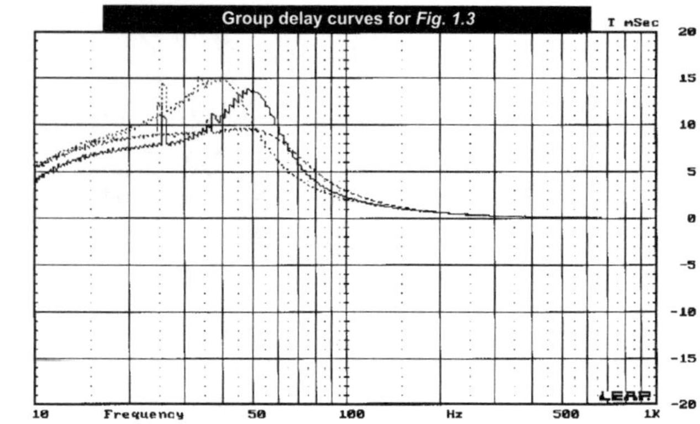

— .53cf 4" port 1 watt
········ .53cf 6" port 1 watt
- - - .34cf (QB3) 6" port 1 watt

FIGURE 1.4

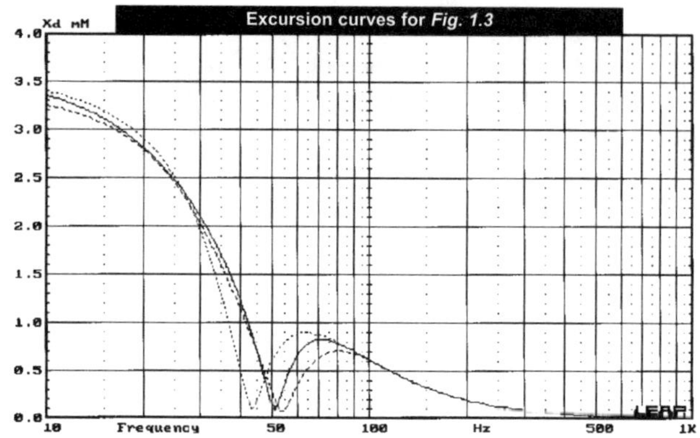

— .53cf 4" port 1 watt
········ .53cf 6" port 1 watt
- - - .34cf (QB3) 6" port 1 watt

FIGURE 1.5

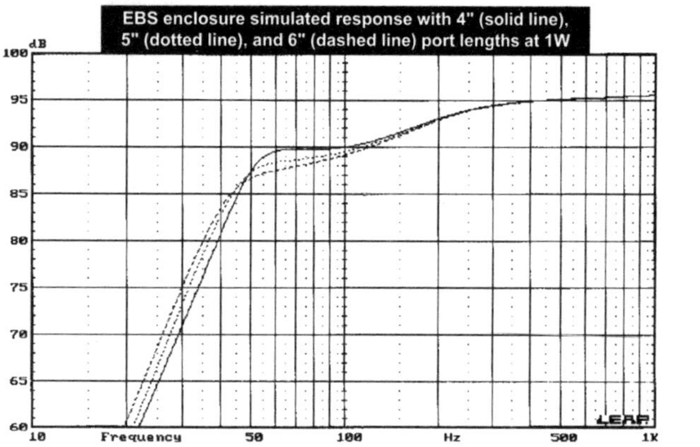

— .53cf 4" port 1 watt
········ .53cf 5" port 1 watt
- - - .53cf 6" port 1 watt

FIGURE 1.6

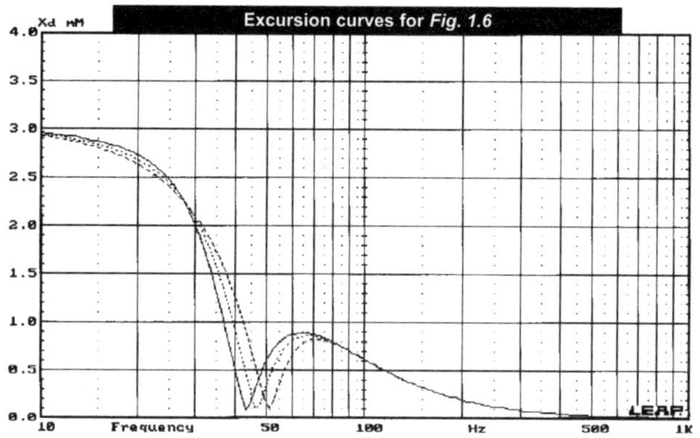

— .53cf 6" port 1 watt
········ .53cf 5" port 1 watt
- - - .53cf 4" port 1 watt

FIGURE 1.7

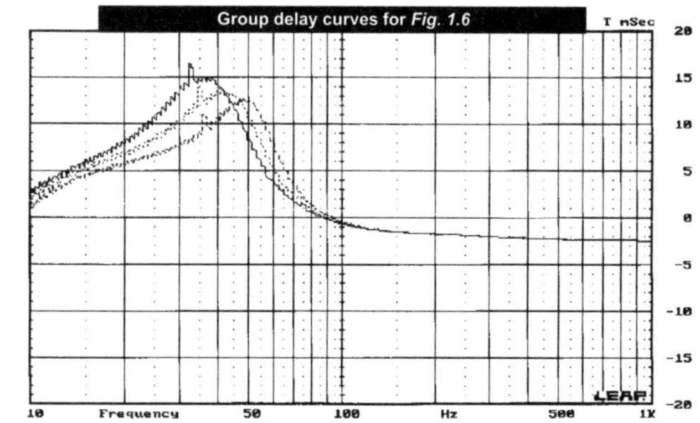

— .53cf 6" port 1 watt
········ .53cf 5" port 1 watt
- - - .53cf 4" port 1 watt

FIGURE 1.8

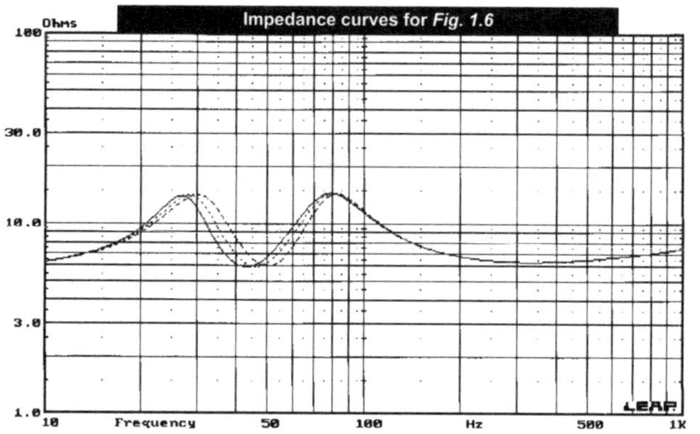

— .53cf 6" port 1 watt
········ .53cf 5" port 1 watt
- - - .53cf 4" port 1 watt

FIGURE 1.9

measurement by subtracting 6dB, and spliced it to a gated-anechoic measurement made on a 5.5' tower. I made the frequency splice somewhere between 300-400Hz. The curve in *Fig. 1.21* shows the woofer with both 4" and 6-inch vent lengths.

The woofer's off-axis response, given in *Fig. 1.22*, is down about 6dB between 0° and 45° at 3kHz, but is down only 2dB at 2kHz. This critical information can help determine the crossover frequency choice.

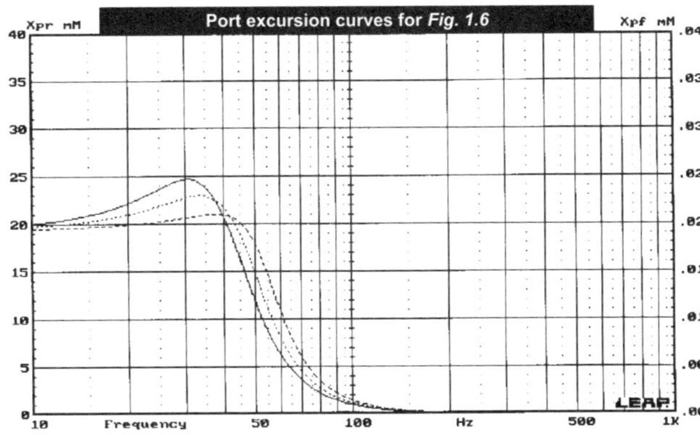

—— .53cf 6" port 1 watt
.... .53cf 5" port 1 watt
- - - .53cf 4" port 1 watt

FIGURE 1.10

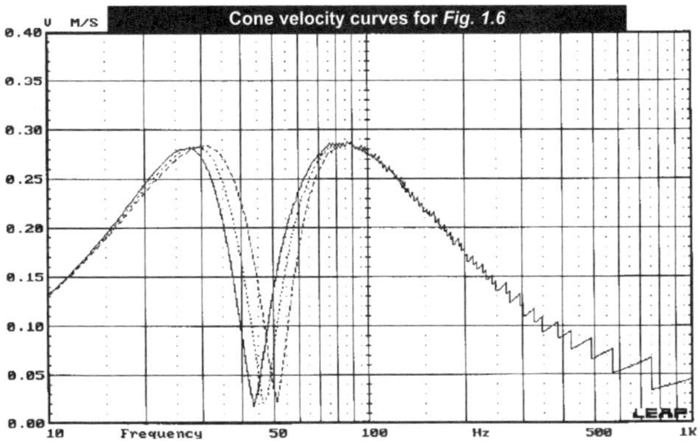

—— .53cf 6" port 1 watt
.... .53cf 5" port 1 watt
- - - .53cf 4" port 1 watt

FIGURE 1.11

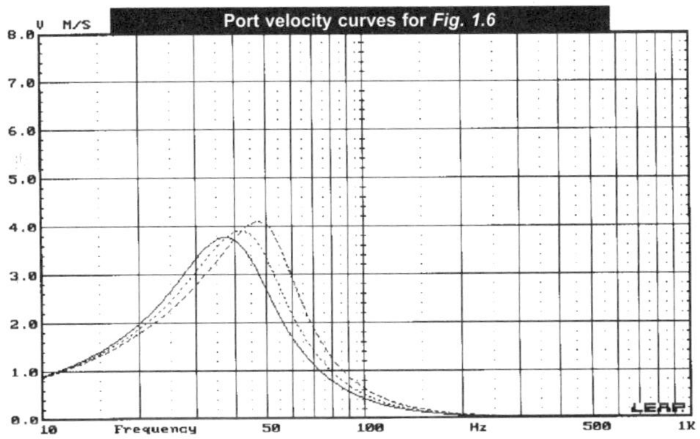

—— .53cf 6" port 1 watt
.... .53cf 5" port 1 watt
- - - .53cf 4" port 1 watt

FIGURE 1.12

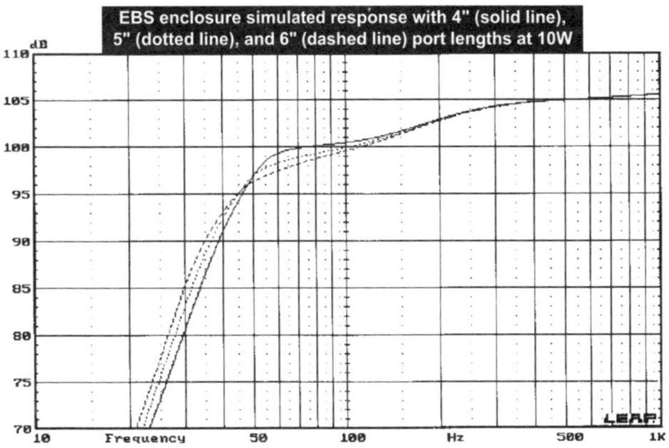

—— .53cf 4" port 10 watts
.... .53cf 5" port 10 watts
- - - .53cf 6" port 10 watts

FIGURE 1.13

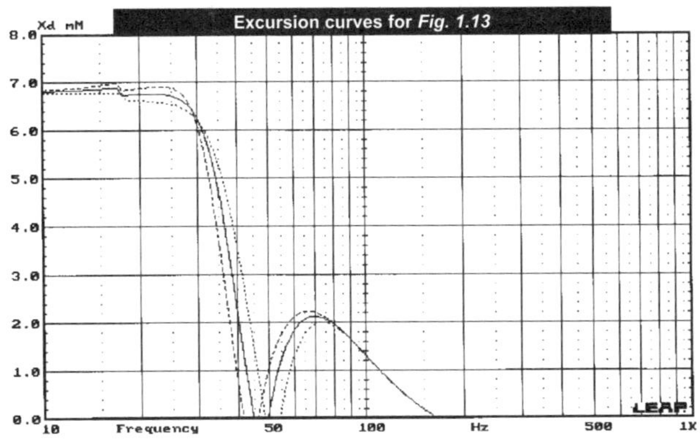

—— .53cf 5" port 10 watts
.... .53cf 4" port 10 watts
- - - .53cf 6" port 10 watts

FIGURE 1.14

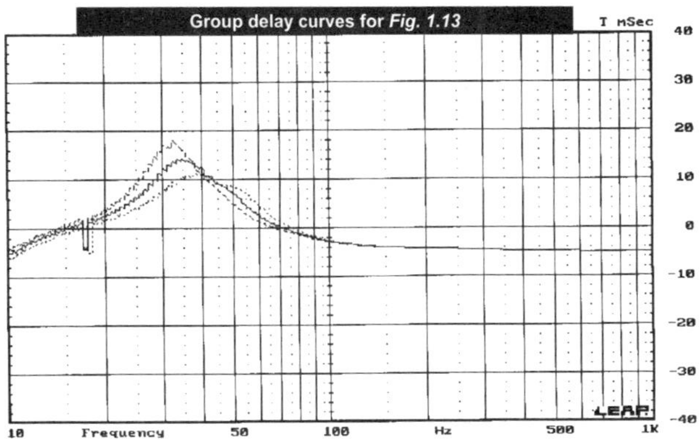

—— .53cf 5" port 10 watts
.... .53cf 4" port 10 watts
- - - .53cf 6" port 10 watts

FIGURE 1.15

One important factor in producing a stable acoustic image is the response matching of the left and right speakers. *Figure 1.23* (made with a 6-inch vent length) compares the full-range anechoic response of the two woofer samples. Variations are no greater than an acceptable 1dB. *Figure 1.24* shows the measured impedance of the speaker, with a 4" vent—ultimately the specified length.

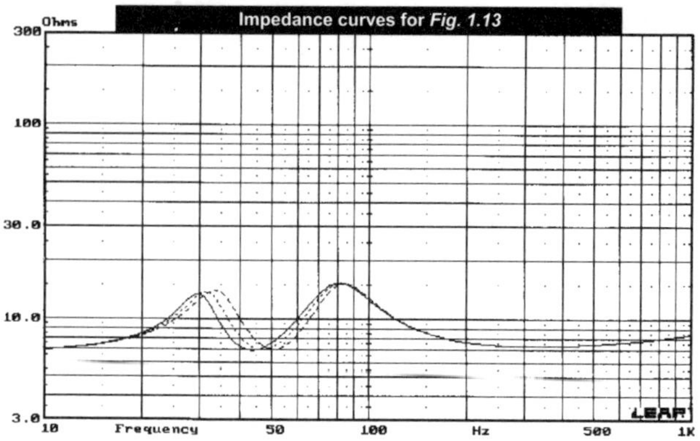

—— .53cf 6" port 10 watts
.... .53cf 5" port 10 watts
- - - .53cf 4" port 10 watts

FIGURE 1.16

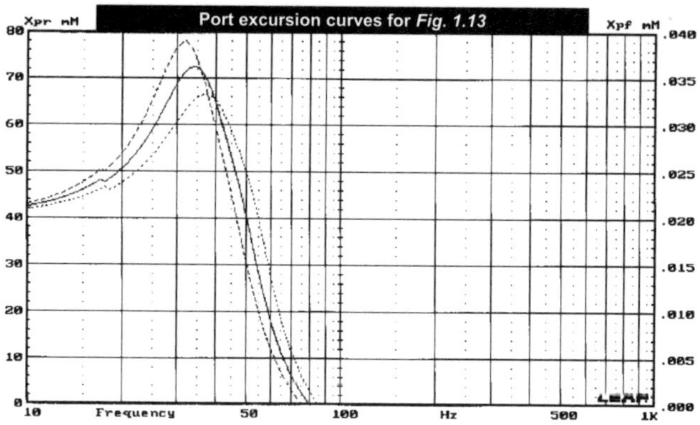

—— .53cf 5" port 10 watts
.... .53cf 4" port 10 watts
- - - .53cf 6" port 10 watts

FIGURE 1.17

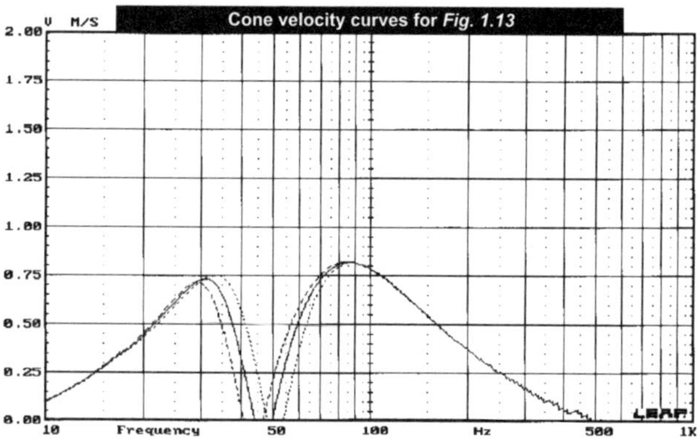

—— .53cf 5" port 10 watts
.... .53cf 4" port 10 watts
- - - .53cf 6" port 10 watts

FIGURE 1.18

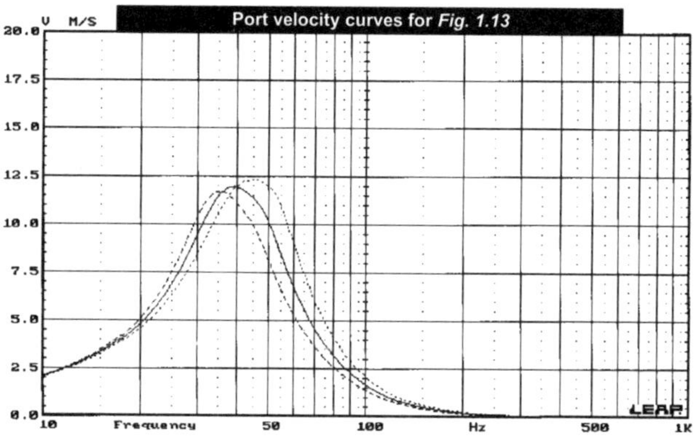

—— .53cf 5" port 10 watts
.... .53cf 4" port 10 watts
- - - .53cf 6" port 10 watts

FIGURE 1.19

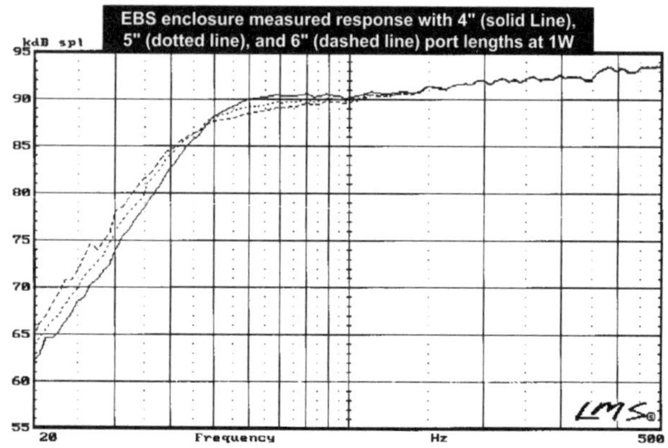

—— qp 4" no xover
.... qp 5" no xover
- - - qp 6" no xover

FIGURE 1.20

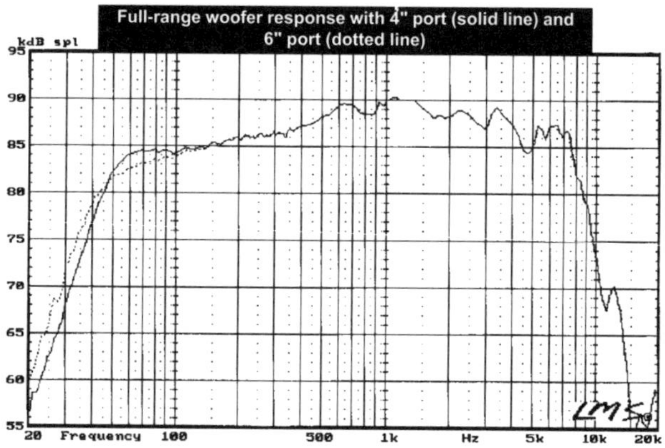

—— Full range 4"
.... Full range 6" A

FIGURE 1.21

1.30 The Tweeter

The Scan-Speak D2905 tweeter is an example of the best in 1-inch soft-dome construction. Released about two years ago, it has undergone continuous refinement. The current version, which Proac uses in some of its more expensive models, has a metal flange, doped 1-inch cloth dome, vented pole piece, and plastic rear cavity.

My first measurement of this tweeter was an impedance sweep, shown in *Fig. 1.25.* Apparently, this version did not include magnetic fluid and had a resonance slightly lower than 1kHz. I transferred the data to LEAP and calculated the parameters for the following specifications:

Z_{NOM}	5Ω
R_E	4.42Ω
f_S	909.6Hz
Q_{MS}	1.090
Q_{ES}	1.112
Q_{TS}	0.55
L_E (1kHz)	0.107mH
(20kHz)	0.050mH

The full-range anechoic frequency response from 0°-45° (*Fig. 1.26*) shows a smooth response of ±2.5dB from 1-18kHz. Both D2905 samples are compared in *Fig. 1.27*, which I expect is typical of what you might get when you order a set. The pair shown varied about 1.35dB below 1kHz, and 1.5dB above 10kHz. They could be better matched if you were able to sort through a quantity of drivers, but are nonetheless acceptable as is.

1.40 Determining the Crossover Configuration

A two-way speaker's typical crossover frequency is 1.8-4kHz, depending on the woofer's cone diameter. As noted earlier, an optimum system off-axis response would occur in the vicinity of 2kHz. Other factors make this desirable, the most important of which concerns the time delay resulting from mounting the two drivers flush on one baffle.

The time (phase) delay between two drivers mounted in this way relates to the physical distance between the initial radiating points. Such concepts as acoustic center, which depicts overall driver group delay are not relevant. In the case of the P17WJ and the D2905, the distance between the two frontplates mounted and inset on the same baffle is 1.63", or 120μs.

The curves in *Figs. 1.28* and *1.29* show the amplitude relationships of electrical in-phase and out-of-phase connections for two theoretically flat drivers. The 120μs delay distance is crossed over with fourth-order Linkwitz-Riley (L-R) network sections at 2.2kHz and 3kHz, respectively. The drivers are nearly 180° out-of-phase at 3kHz (*Fig. 1.29*), as indicated by the deep 8dB in-phase null at 3.1kHz. The out-of-phase connection is down only 2dB at 2kHz. Both polarities produce a 3+dB null at 2.2kHz, in or out of phase due to the shorter wavelength, indicating that less than 180° phase difference exists at that frequency (*Fig. 1.28*).

Altering this time delay by 50%, to 60μs, changes the picture dramatically (*Figs 1.30* and *1.31*). The in-phase curve at 3kHz in *Fig. 1.31* is down 2dB, while the one at 2.2kHz is down by only about 1dB, providing a usable solution.

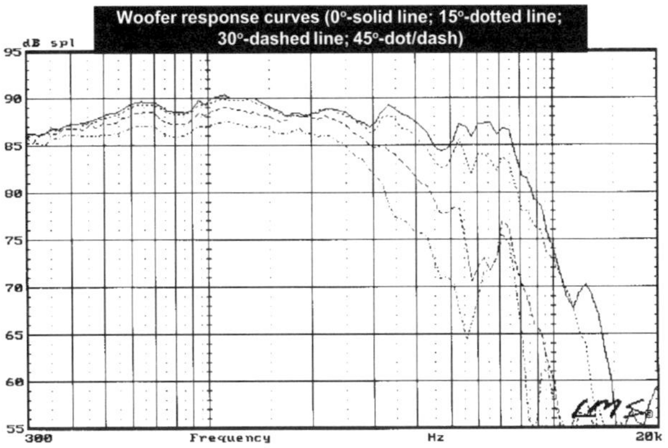

—— Vifa PW17 0
.... Vifa PW17 15
- - - Vifa PW17 30
-.-. Vifa PW17 45

FIGURE 1.22

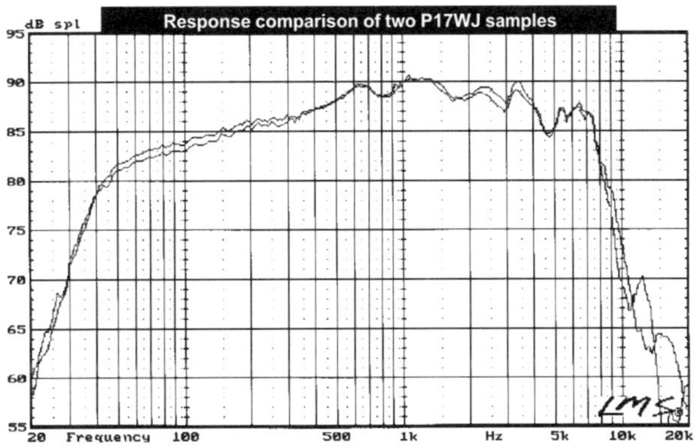

—— Full range 6" A
- - - Full range 6" B

FIGURE 1.23

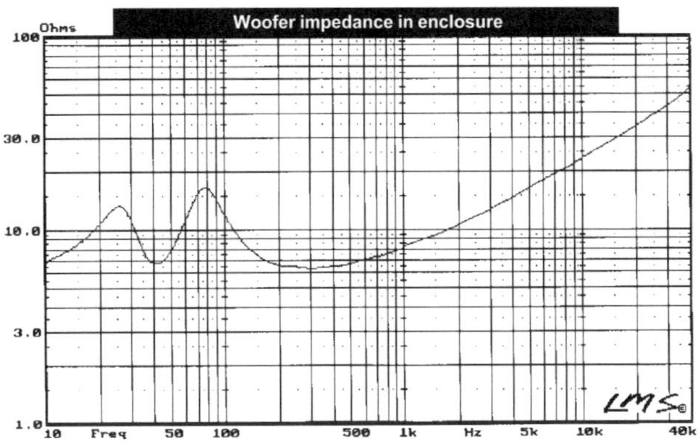

.... Woofer impedance

FIGURE 1.24

The 2.2kHz crossover frequency is preferable if you can limit the time delay between the two drivers to 60μs. Each speaker in this book uses a different technique to accomplish this task and, in this case, I chose to reduce the delay difference by tilting the speaker back. The tweeter, mounted above the

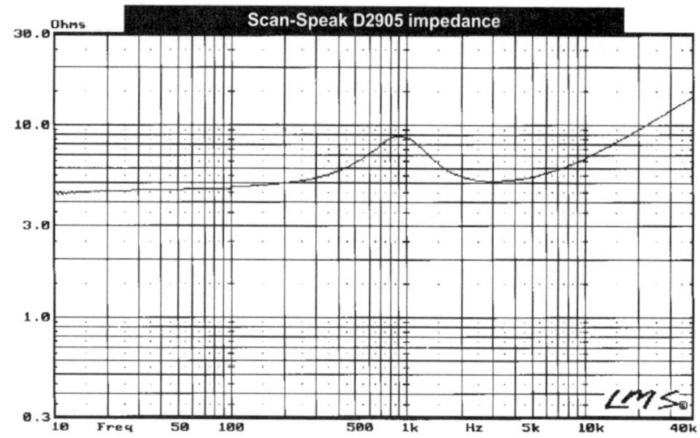

-·-· D2905 impedance

FIGURE 1.25

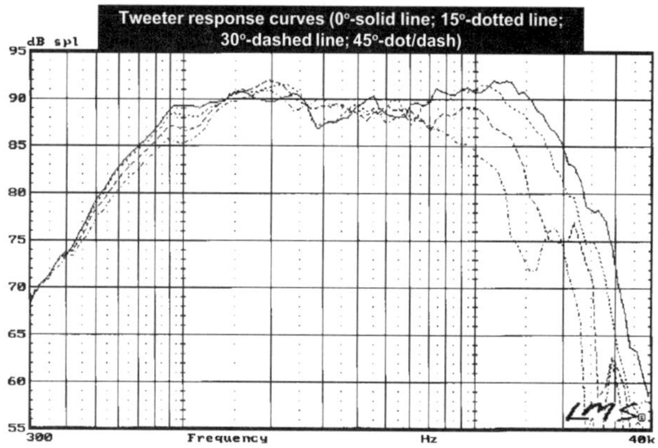

—— D2905 0
········ tw15
- - - tw30
-·-· tw45

FIGURE 1.26

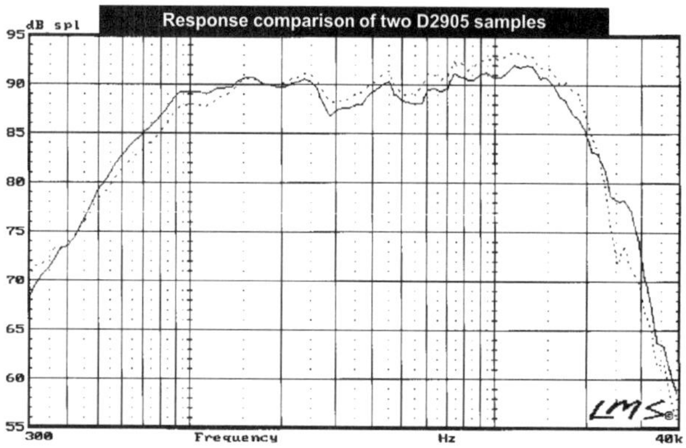

—— D2905 sample a
- - - D2905 sample b

FIGURE 1.27

woofer, will be closer to the woofer's radiating plane.

Tilting the speaker or sloping the front baffle works well, although it's not the same as using a purely electrical delay. A sloped baffle places the drivers substantially off-axis at the listening position. You can probably remove the entire 120μs of delay by tilting the speaker, but the required angle would alter the frequency response too much by placing the speaker farther off-axis. Using trial-and-error and a measurement technique I'll explain later, I determined that tilting the speaker back only 6° will reduce the time difference to 60μs.

You can easily measure the time delay with a ruler or tape measure. For the P17WJ, the distance from the front baffle (top of the basket) to the center of the magnet assembly's front plate is 1.75"; the distance is 0.1875" for the D2905. Subtracting the two measurements leaves 1.5625", or 116μs (the time it takes sound to travel 1.56"). I confirmed this using LMS and a procedure I described in the October 1992 issue of *Voice Coil.*

The process is fairly simple: you begin by measuring the two drivers' full-range response using a crossover. While it need not be designed specifically for the drivers being tested, the crossover should be at least a second-order topography. I use a basic two-way network with a 2kHz crosspoint and a fourth-order topography on the high-pass section, and a first-order topography and conjugate on the low-pass. For the purposes of this test, it does not matter whether the resulting frequency response is flat or has dips and peaks—only that each speaker's acoustic slopes be at least second-order to minimize the overlap between high- and low-pass sections. The two network sections must also be separated so you can test the woofer and tweeter either individually or together.

After making the full-range measurement using the crossover, measure each section separately: the woofer with the low-pass and the tweeter with the high-pass network. The trick is to compare the mathematical summation of the individual measurements with the measured summation of the two drivers.

Figure 1.32 shows this comparison. The mathematical summation varies from the measured summation (note the large dip in the response) because, as yet, nothing has been done to account for the physical delay between the drivers. (The trial crossover measured fairly flat only by accident; the crossover I used was designed for a double-driver configuration using two entirely different speakers.) To finish the procedure, add the various low-pass delay values to the woofer curve individually. Compare each result to the measured summation until you find a curve that overlaps it.

Figure 1.33 shows the result using the LMS Time Delay command to add 130μs of delay to the woofer curve before combining it with the tweeter curve. As you can see, the mathematically summed curve is mostly below the measured curve. With a 100μs delay (*Fig. 1.34*) the mathematically summed curve lies primarily above the measured

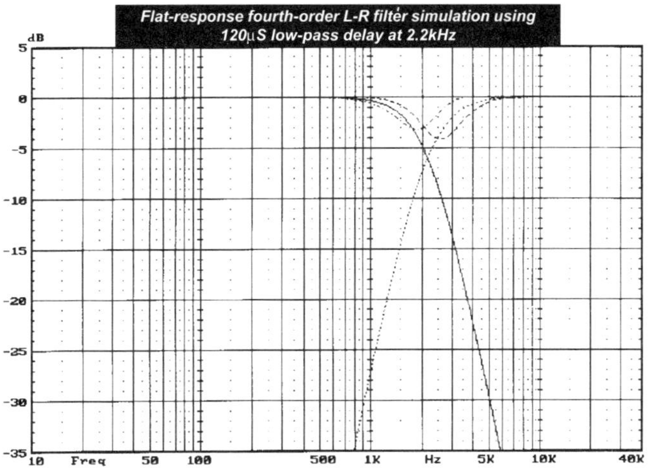

FIGURE 1.28

—— Lo Pass Section Only: Magnitude
········ Hi Pass Section Only: Magnitude
– – – Sum= Lo+Hi (InPhase): Magnitude
–·–· Sum= Lo-Hi (Reverse): Magnitude

Lo Filtr: 4th-LnkRly Hi Filtr: 4th-LnkRly
Freq = 2200.00 Hz Freq= 2200.00 Hz
Level= 0.00 dB Level= 0.00 dB
Delay= 120.00µS Delay= 0.00 µS

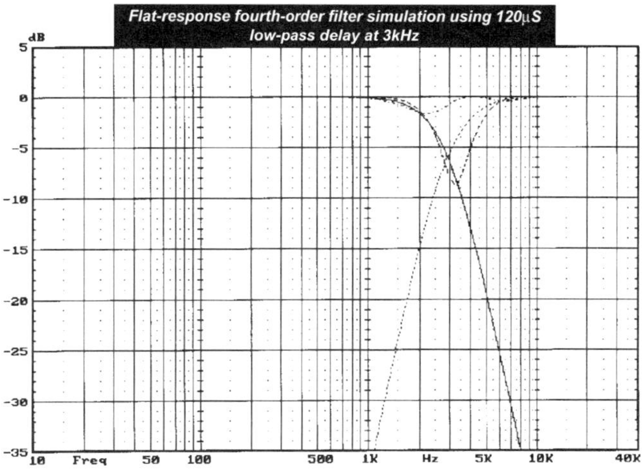

FIGURE 1.29

—— Lo Pass Section Only: Magnitude
········ Hi Pass Section Only: Magnitude
– – – Sum= Lo+Hi (InPhase): Magnitude
–·–· Sum= Lo-Hi (Reverse): Magnitude

Lo Filtr: 4th-LnkRly Hi Filtr: 4th-LnkRly
Freq = 3000.00 Hz Freq= 3000.00 Hz
Level= 0.00 dB Level= 0.00 dB
Delay= 120.00µS Delay= 0.00 µS

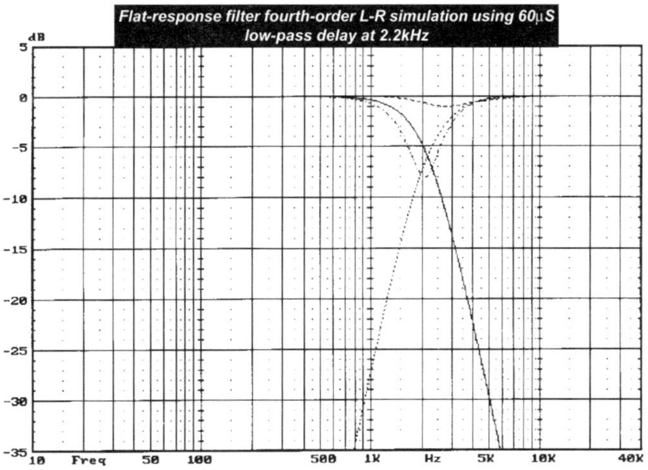

FIGURE 1.30

—— Lo Pass Section Only: Magnitude
········ Hi Pass Section Only: Magnitude
– – – Sum= Lo+Hi (InPhase): Magnitude
–·–· Sum= Lo-Hi (Reverse): Magnitude

Lo Filtr: 4th-LnkRly Hi Filtr: 4th-LnkRly
Freq = 2200.00 Hz Freq= 2200.00 Hz
Level= 0.00 dB Level= 0.00 dB
Delay= 60.00µS Delay= 0.00 µS

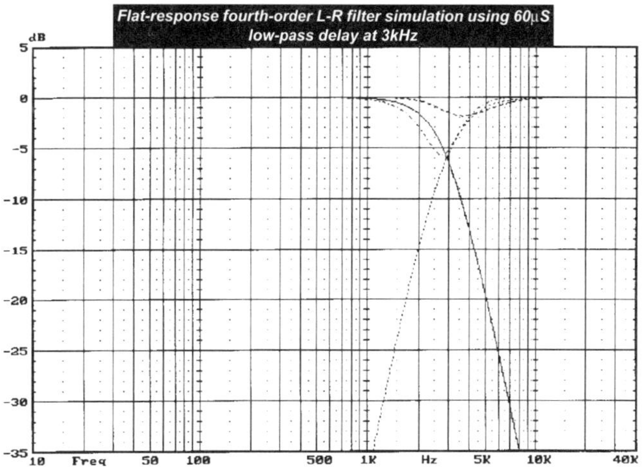

FIGURE 1.31

—— Lo Pass Section Only: Magnitude
········ Hi Pass Section Only: Magnitude
– – – Sum= Lo+Hi (InPhase): Magnitude
–·–· Sum= Lo-Hi (Reverse): Magnitude

Lo Filtr: 4th-LnkRly Hi Filtr: 4th-LnkRly
Freq = 3000.00 Hz Freq= 3000.00 Hz
Level= 0.00 dB Level= 0.00 dB
Delay= 60.00µS Delay= 0.00 µS

summation. The result of adding 120µs of delay (within 4µs of the ruler measurement) is shown in *Fig. 1.35.* The curves nearly overlap, indicating the simulated 120µs delay is very close to the real delay. This measurement is not critical, and as far as computer-aided engineering (CAE) of crossovers is concerned, accuracy to within 10µs or so is more than sufficient.

By using the same technique and measuring the delay at different baffle angles, I easily determined that 6° would provide the needed 60µs of delay. *Figures 1.36*, *1.37*, and *1.38* illustrate the same process using 50µs, 60µs, and 70µs of delay. The delay amount measured at the on-axis position (with the microphone placed at 1m between two drivers) was very close to 60µs. By using a 2.2kHz

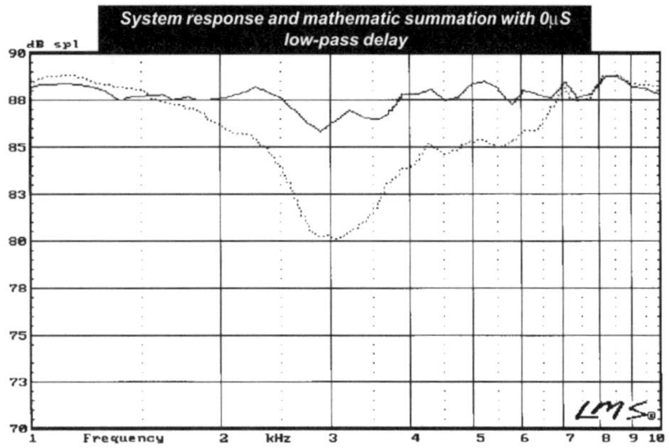

— Sys 0 sw
······· Sum 0 0 delay

FIGURE 1.32

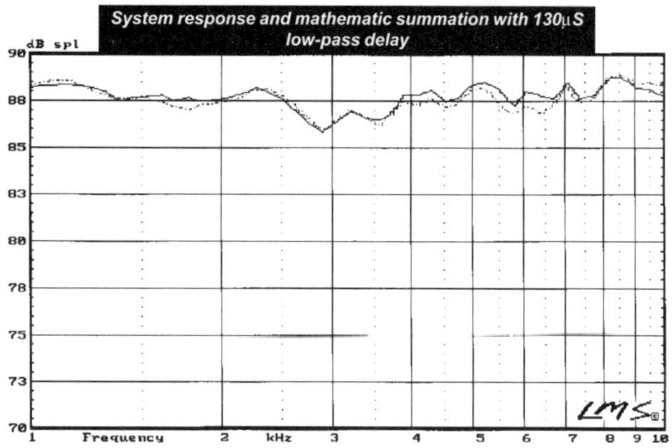

— Sys 0 sw
-·-· Sum 0 130μS

FIGURE 1.33

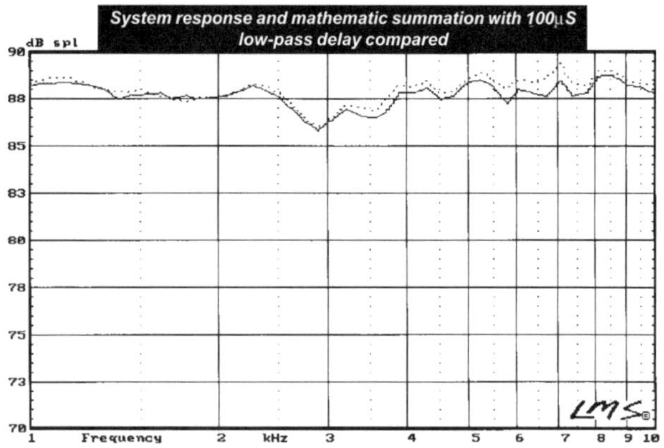

— Sys 0 sw
- - - Sum 0 100μS

FIGURE 1.34

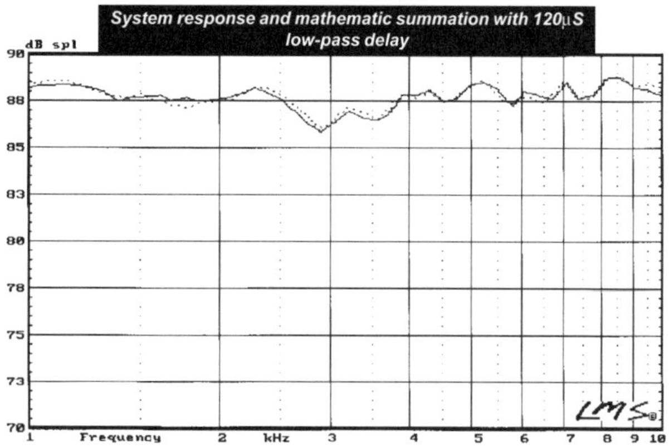

— Sys 0 sw
- - - Sum 0 120μS

FIGURE 1.35

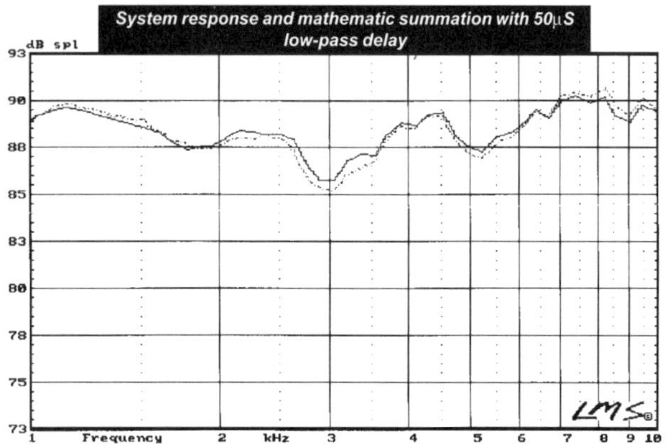

— Sys sw
-·-· Sum 50μS delay

FIGURE 1.36

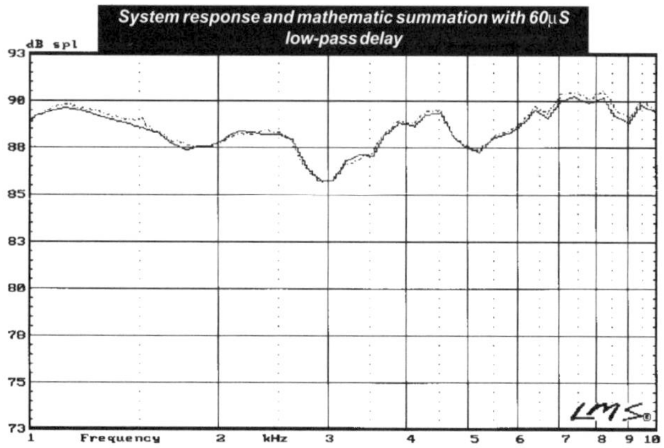

— Sys sw
-·-· Sum 60μS delay

FIGURE 1.37

crossover frequency and 6° of tilt you can optimize both driver networks for a fourth-order Linkwitz-Riley target response. The result should be a flat magnitude response throughout the crossover region.

1.50 The Woofer Crossover

You can use many different crossover topographies with the same two drivers to produce an unlimited variety of flat speakers. If you were to prototype several of these variations, you would discover that

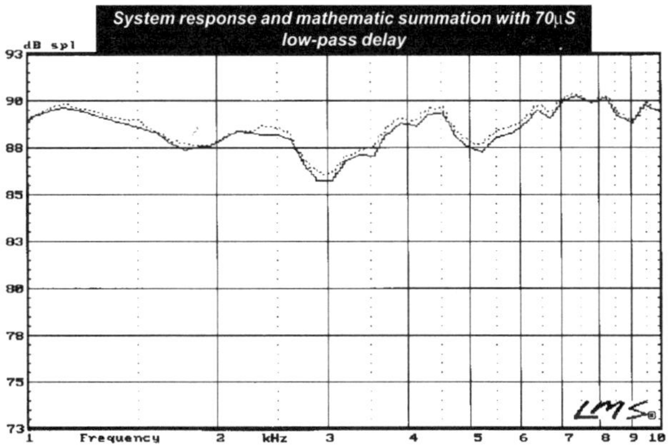

— Sys sw
······· Sum 70µS delay

FIGURE 1.38

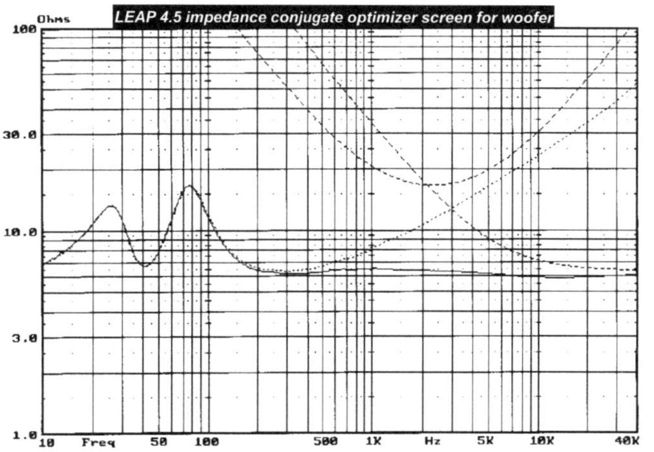

— Combined Impedance
······· Original Impedance
- - - Conjugate Nets

DGL Data Info
DGL# 2:P17WJ woofer

FIGURE 1.39

Crossover schematic of second-order low-pass topography

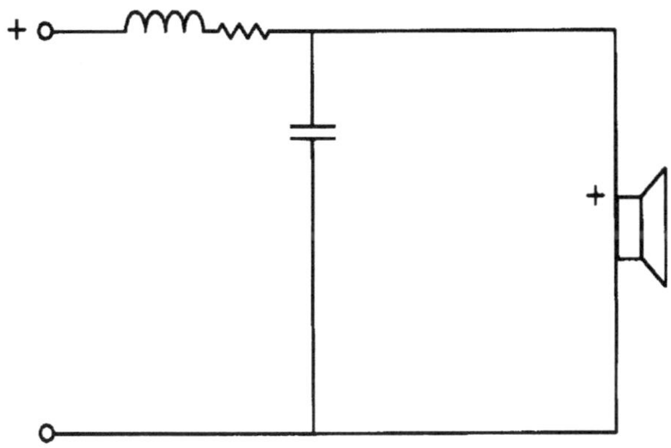

FIGURE 1.40

Crossover schematic of second-order low-pass topography plus LCR and CR conjugate circuits

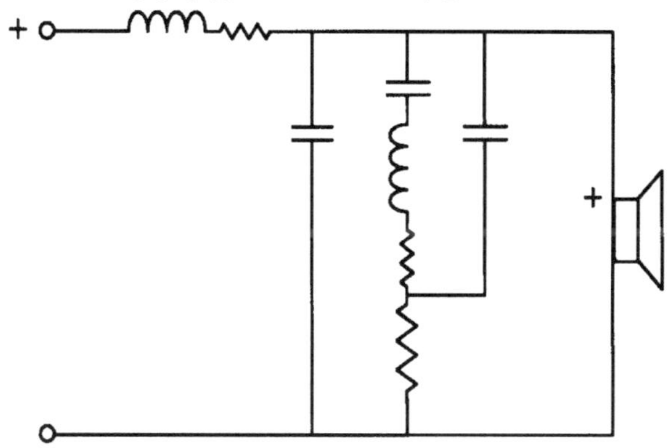

FIGURE 1.41

not all of them sound alike, and not all sound good. While equal SPL for all frequencies is generally associated with good sound, meeting that "flatness" criterion does not ensure good results. One goal of this book is to illustrate that phenomenon. I will therefore discuss each of the four speakers in terms of several alternate crossover networks and also explain how to determine which is more appropriate.

Figure 1.21 shows the woofer response rising from about 84dB at 100Hz to 90dB at 1kHz, and then dropping back to 88dB at the 2.2kHz target crossover frequency. The rise is caused by cabinet diffraction, which, because of the front baffle's size and dimensions, reflects sound better at high frequencies than low. The crossover must begin to alter the woofer's response at around 200Hz, and cause the voltage drive to decrease in amplitude at the same rate the woofer rises in acoustic amplitude. The crossover must then begin to attenuate the response at 24dB/octave above frequency.

One way to design a speaker crossover network is to first produce a flat impedance response using conjugate networks. The LEAP software can readily create initial values for these types of circuits. *Figure 1.39* shows the LEAP Conjugate Designer utility screen after calculating an inductor, capacitor, resistor (LCR) series conjugate at 2,373Hz, and capacitor, resistor (CR) conjugate to counteract the P17WJ's reactive impedance rise. The values for the two circuits were:

LCR	CR
0.42mH	4.75µF
10.59µF	6.34Ω
16.59Ω	

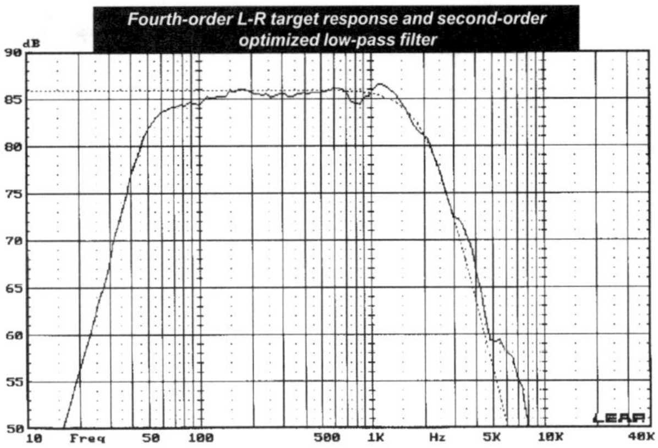

—— Wof 1 4" port p1
.... lp target

FIGURE 1.42

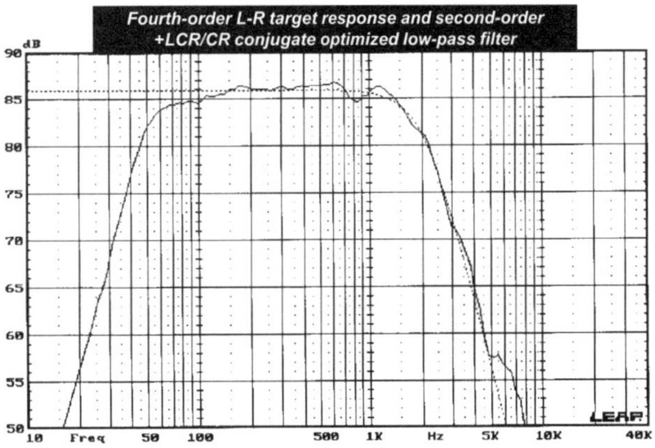

—— Wof 2a 4" port p2
.... lp target

FIGURE 1.43

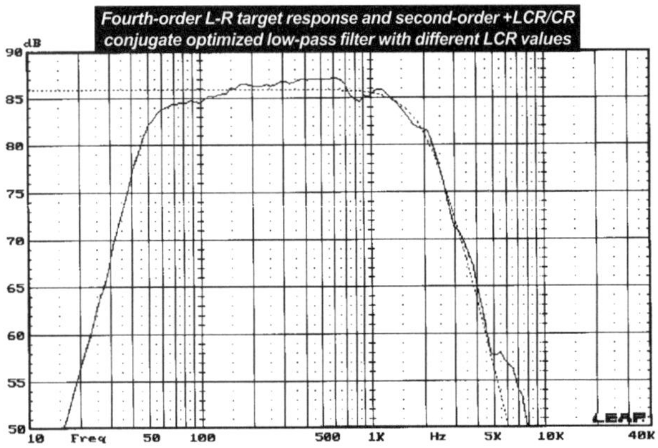

—— Wof 2b 4" port p3
.... lp target

FIGURE 1.44

Since these were initial approximations, they were not adjusted to reflect standard component values. While this can be a viable method for beginning a design, employing a computer optimizer seldom results in using a conjugate specifically tailored for a flat impedance.

The three low-pass solutions I arrived at for the P17WJ used two basic schematic topographies, as illustrated in *Figs. 1.40* and *1.41* (the small resistors represent the direct current resistance (DCR) of the inductors). *Figure 1.40* represents a basic second-order configuration; the other two solutions used a second-order plus an LCR and CR conjugate (*Fig. 1.41*). Both LCR and CR employed a single resistor.

Describing the computer optimization procedure I used to derive the crossover values would be tedious and boring so I will concentrate on evaluating the results once the target response is achieved. I should emphasize that network values derived from formulas dependent upon resistive termination when applied to typically reactive loudspeaker loads rarely produce expected results. In practice, only trial and error or sophisticated computer software produce accurate network values.

Figures 1.42, *1.43*, and *1.44* depict the final optimized response of the three low-pass crossover solutions. *Figure 1.42*, the first circuit, is the second-order layout; *Figs. 1.43* and *1.44*, (the second and third circuits, respectively) combine second-order and conjugate layouts. The dashed line represents a fourth-order L-R low-pass target response at 2.2kHz referenced to 86dB. I designated this as the target SPL so the bottom of the speaker would fall off somewhat. A droop in the low-end response occurs at 100Hz, a region where most small-size rooms begin to experience a "lift" in the frequency response (approximately +3dB at 80Hz, increasing to +9dB at 20Hz). To make this speaker perfectly flat starting at the knee of the low-end rolloff would require contouring the entire unit to 83-84dB. I opted for an extra 2dB of overall efficiency in the 100Hz range.

The associated woofer plus the network impedance curves of the first circuit are shown in *Figs. 1.45-1.47*, their transfer functions are given in *Figs. 1.48-1.50*. You can make several observations from the three sets of curves. The simple second-order topography works fairly well in that it fits the target knee (*Figs. 1.42*, *1.45*, and *1.48*). It begins to deviate at about 12dB down below the 86dB SPL, however, the response need not follow the target in its entirety. About 15-20dB down from the target SPL is desirable, with greater variance having little effect on the overall frequency response. Notice also the transfer function of this circuit (and of the other two) has multiple "break points" at about 200Hz and 2kHz, rather than the single knee of a classical filter function.

The other two circuits (*Figs. 1.43*, *1.46*, and *1.49*; and *Figs. 1.44*, *1.47*, and *1.50*) follow the target response in somewhat different ways, although both exceed 25dB down below the 86dB

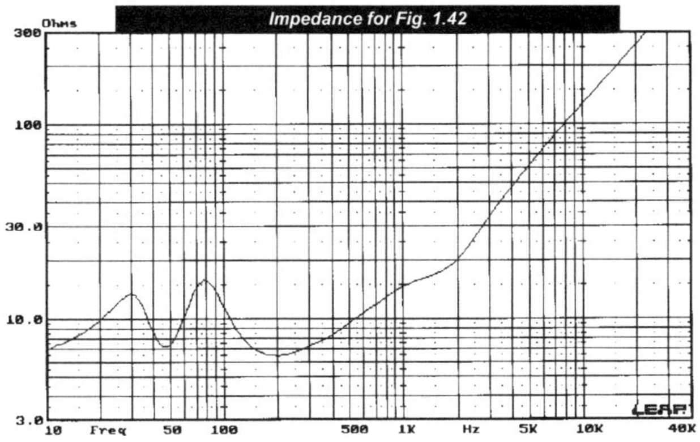

— Wof 1 4" port p1

FIGURE 1.45

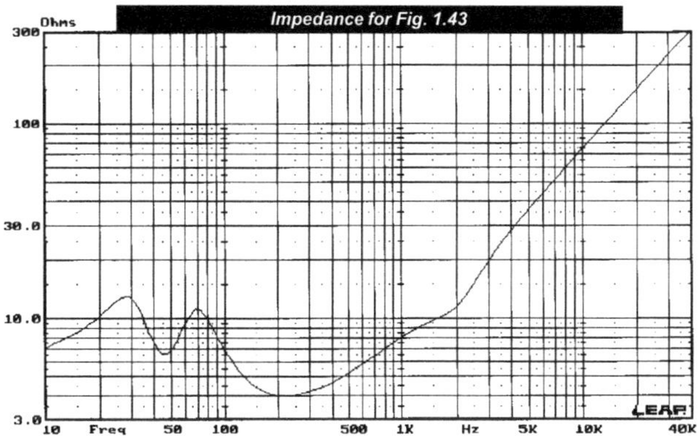

— Wof 2a 4" port p2

FIGURE 1.46

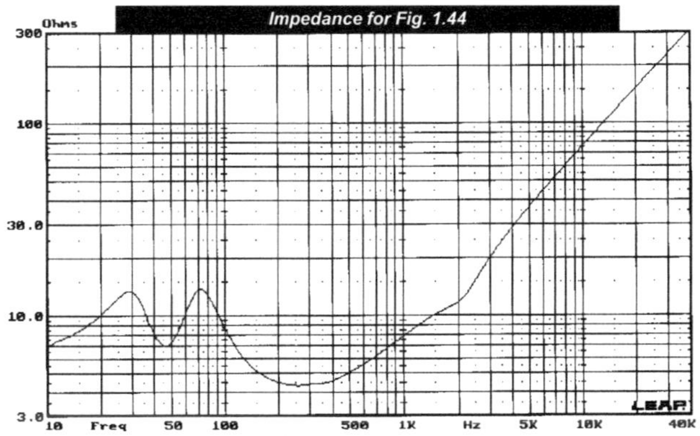

— Wof 2b 4" port p3

FIGURE 1.47

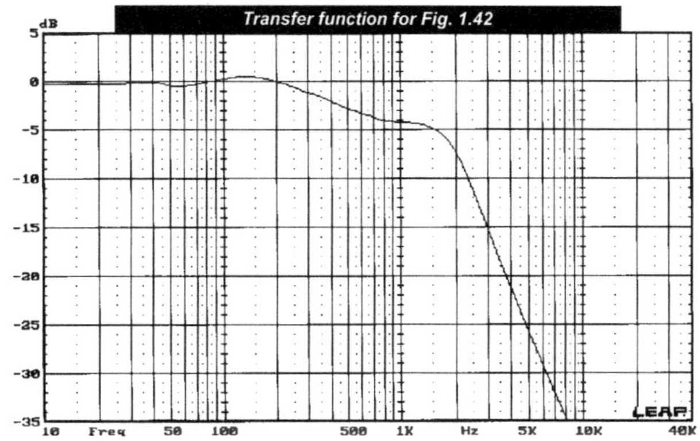

— Wof 1 4" port p1

FIGURE 1.48

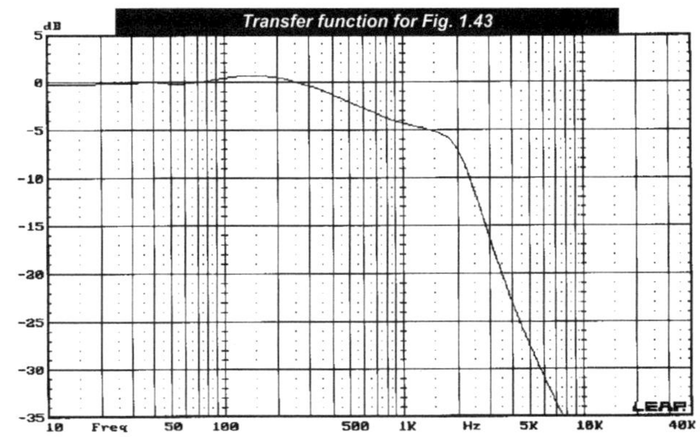

— Wof 2a 4" port p2

FIGURE 1.49

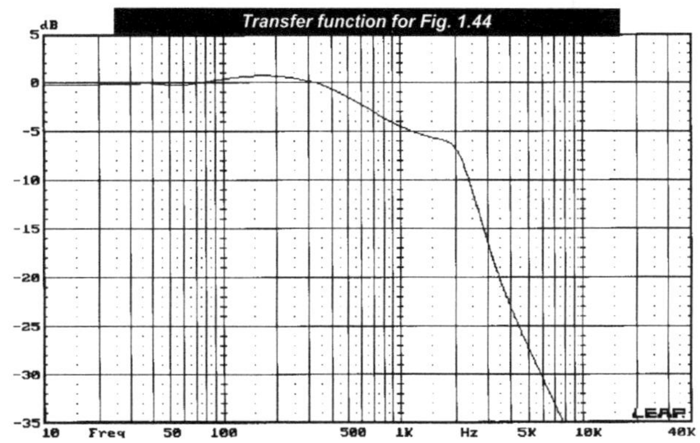

— Wof 2b 4" port p3

FIGURE 1.50

target SPL before deviating substantially. The main difference between them is in deviation from the target curve at the knee between 1kHz and 2kHz, and in the impedance. The second circuit has the lowest impedance minimum at 4Ω.

Figures 1.51-1.53 compare the SPL, impedance, and transfer functions. While the first circuit would probably have been adequate, I chose the third because it better fits the target curve through the knee down to -20dB. It also has a substantially smaller series inductance and accompanying series resistance. *Figures 1.54* and *1.55* compare the final low-pass circuit and the response with and without a network—for both on-axis and 30° off-axis.

1.60 The Tweeter Crossover

I began designing the tweeter network in the same manner as for the woofer—by first examining the

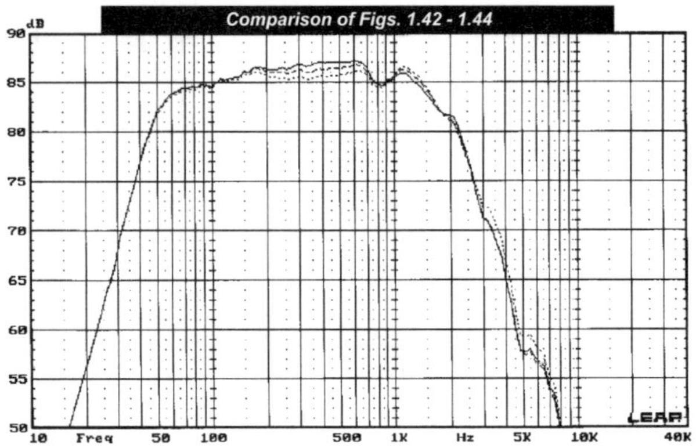

—— Wof 2b 4" port p3
.... Wof 1 4" port p1
– – – Wof 2a 4" port p2

FIGURE 1.51

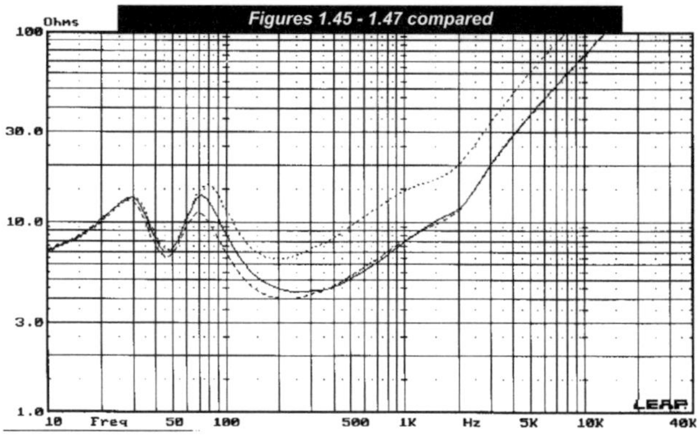

—— Wof 2b 4" port p3
.... Wof 1 4" port p1
– – – Wof 2a 4" port p2

FIGURE 1.52

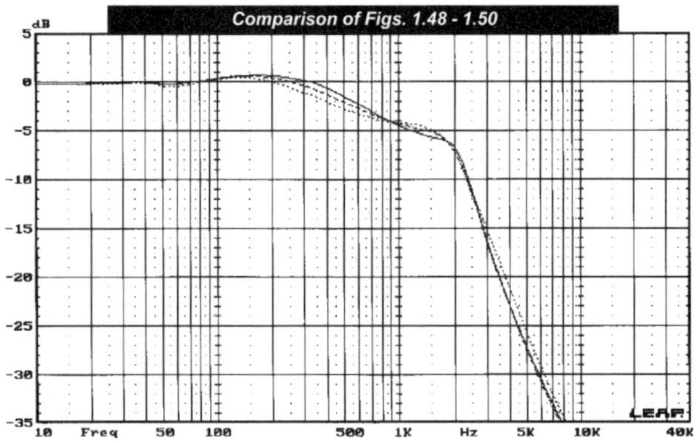

—— Wof 2b 4" port p3
.... Wof 1 4" port p1
– – – Wof 2a 4" port p2

FIGURE 1.53

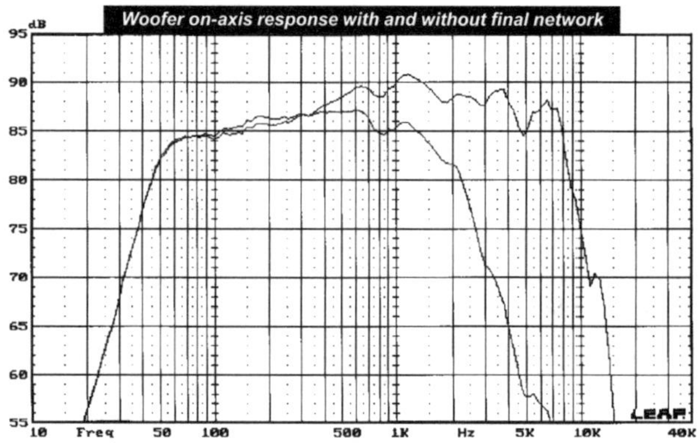

—— Wof 2b 4" port p3
.... P17WJ woofer 4" port

FIGURE 1.54

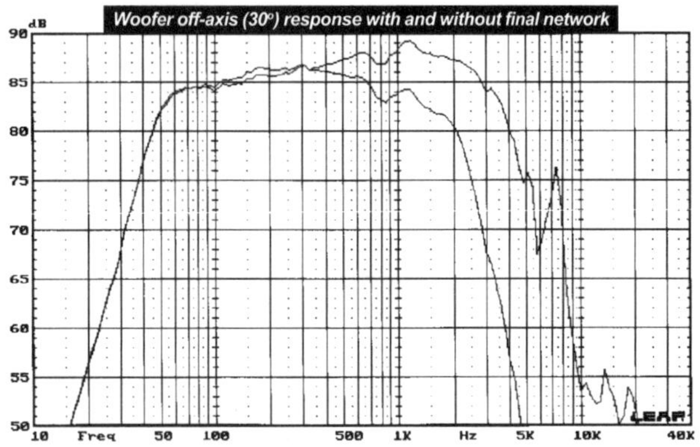

—— Wof 2b 4" port p3
.... P17WJ woofer 4" port

FIGURE 1.55

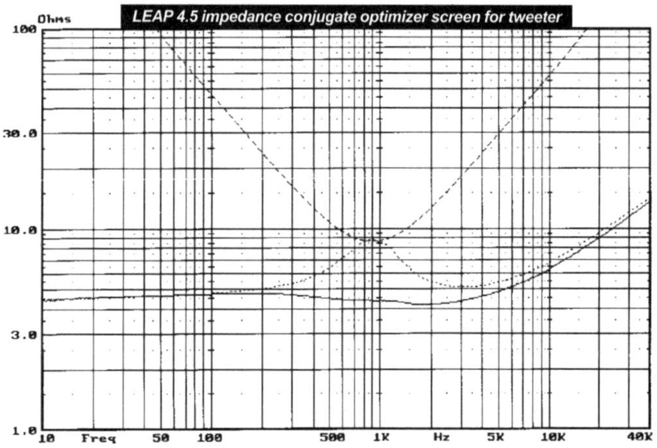

—— Combined Impedance
.... Original Impedance
– – – Conjugate Nets

DGL Data Info
DGL# 9:D2905
new no xover

FIGURE 1.56

values required for an impedance conjugate (in this case, to remove resonance). *Figure 1.56* gives the D2905 impedance curves with and without an LCR conjugate to remove the 907Hz, 0.6 Q resonance. The required values for that circuit are: 0.89mH, 34.4μF, 8.5Ω.

Tweeter circuit #1 using optimized second-order high-pass + LCR conjugate and series attenuation resistor

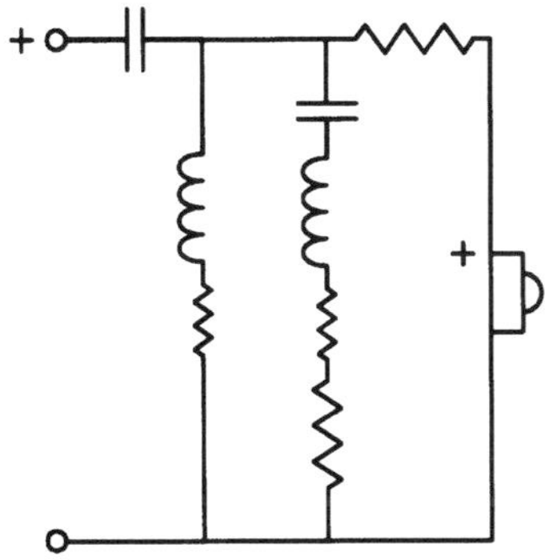

FIGURE 1.57

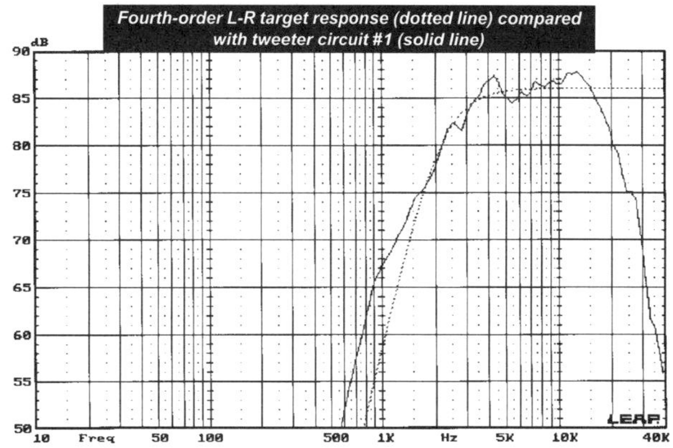

— Tw #1 2nd + conj p4
.... Target

FIGURE 1.58

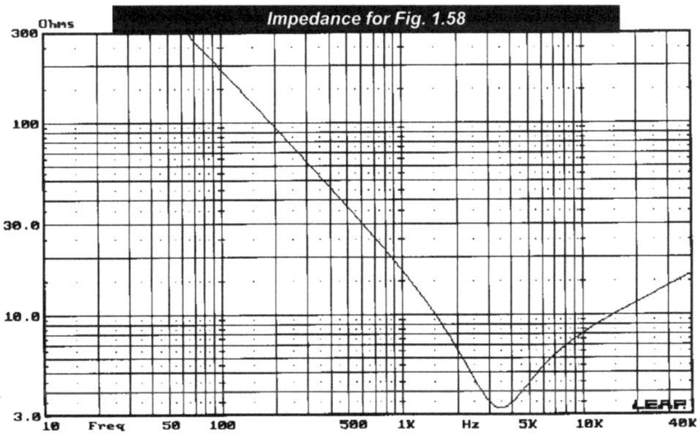

— Tw #1 2nd + conj p4

FIGURE 1.59

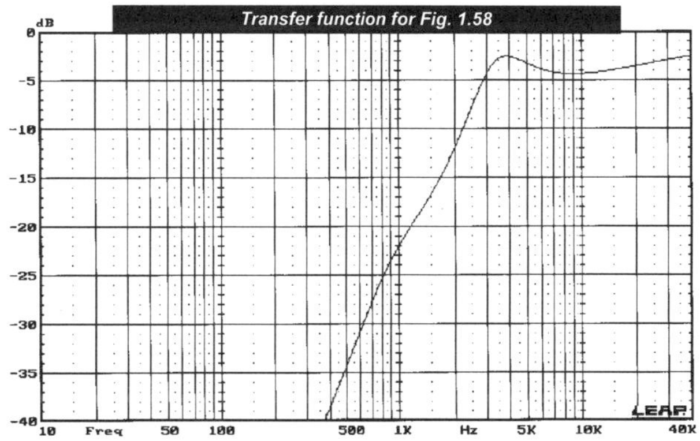

— Tw #1 2nd + conj p4

FIGURE 1.60

Tweeter circuit #2 third-order high-pass + L-pad attenuation resistor network

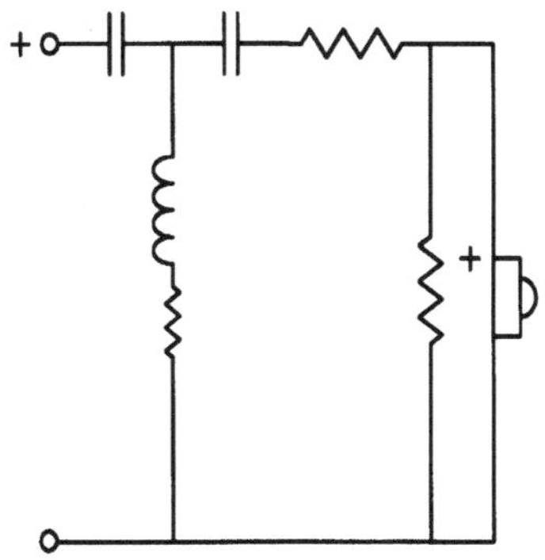

FIGURE 1.61

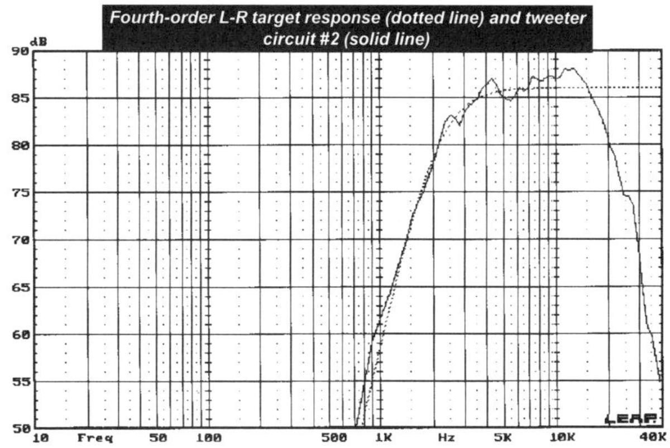

— Tw 3rd + lpad p9
.... Target

FIGURE 1.62

From its frequency response (*Fig. 1.26*) you can see that the D2905 performs down to about 1kHz, where it begins an approximate second-order rolloff. Since this speaker's target SPL is 86dB, the tweeter must also be attenuated by about 5dB. I computer optimized five circuit combinations, using second- and third-order high-pass network topographies, to a fourth-order L-R high-pass target response at 2.2kHz. These circuits are:

1. Second-order high-pass + LCR conjugate + series attenuation resistor.

2. Third-order high-pass + L-pad attenuation circuit.

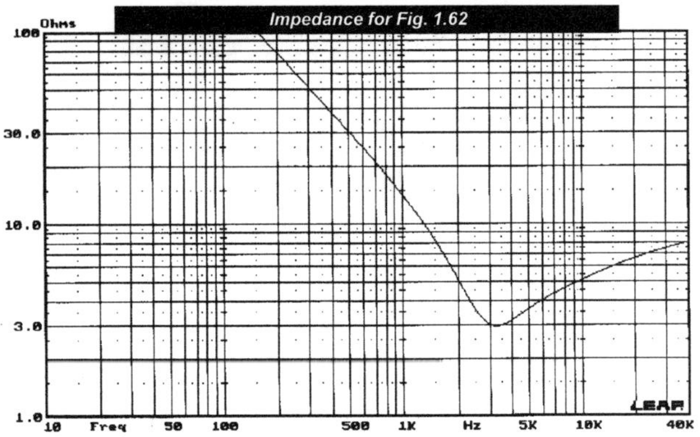

— Tw 3rd + 1pad p9

FIGURE 1.63

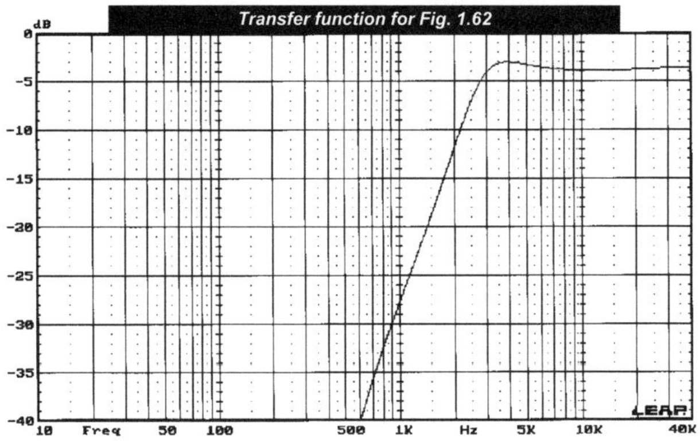

— Tw 3rd + 1pad p9

FIGURE 1.64

Tweeter circuit #3 using third-order high-pass + series attenuation resistor and CR conjugate (switchable).

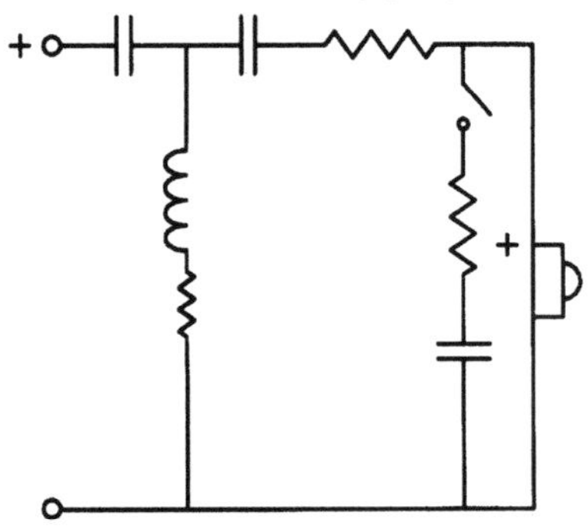

FIGURE 1.65

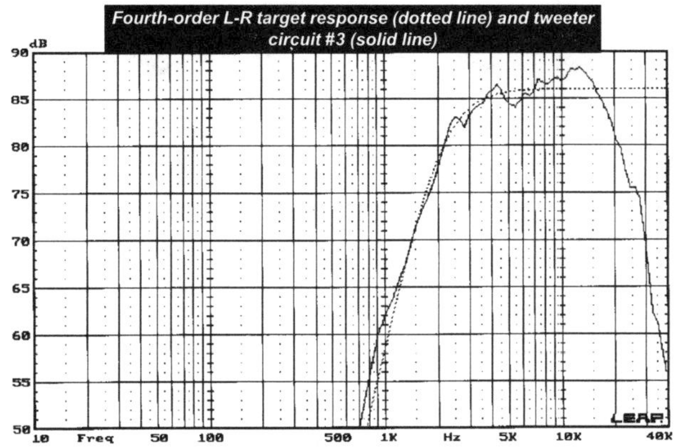

— Tw #2 3rd 2 no cap p5

········ Target

FIGURE 1.66

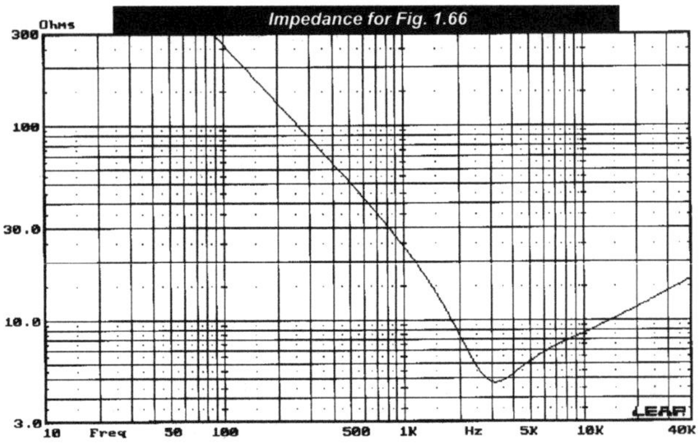

— Tw #2 3rd 2 no cap p5

FIGURE 1.67

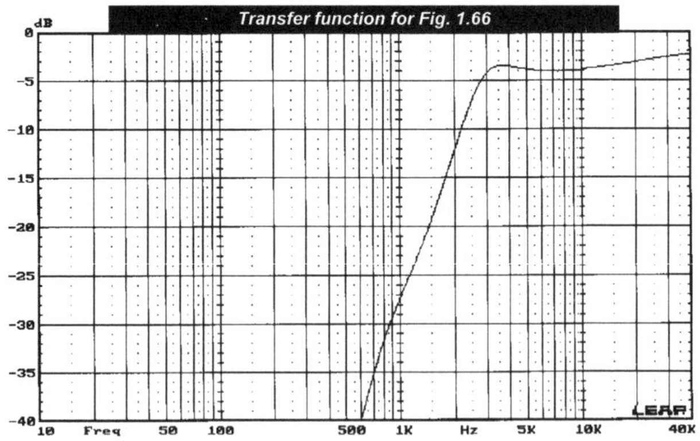

— Tw #2 3rd 2 no cap p5

FIGURE 1.68

3. Third-order high-pass + series resistor + CR conjugate.

4. Same as 3 + capacitor conjugate.

5. Third-order + capacitor conjugate + series resistor + LCR conjugate.

The following diagrams illustrate the five tweeter circuits:

Tweeter #	Circuit	Response +Target	Z	Transfer Function
1	1.57	1.58	1.59	1.60
2	1.61	1.62	1.63	1.64
3	1.65	1.66	1.67	1.68
4	1.69	1.70	1.71	1.72
5	1.73	1.74	1.75	1.76

Tweeter circuit #4 using third-order high-pass + series attenuation resistor and conjugate capacitor

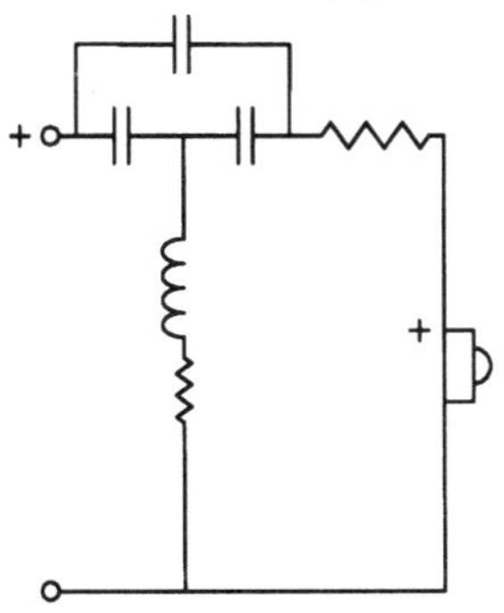

FIGURE 1.69

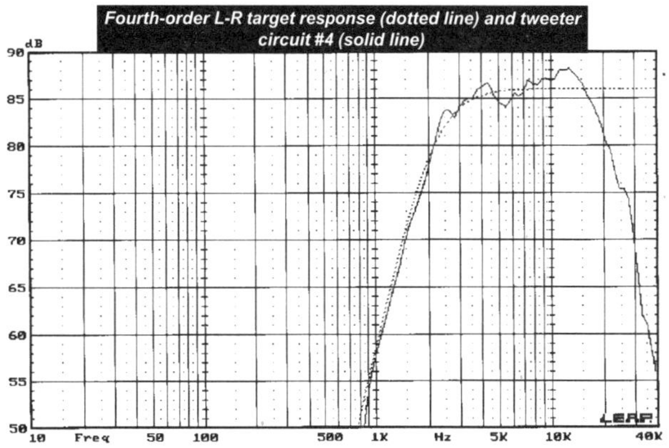

—— Tw #2 3rd 2 with cap p7
.... Target

FIGURE 1.70

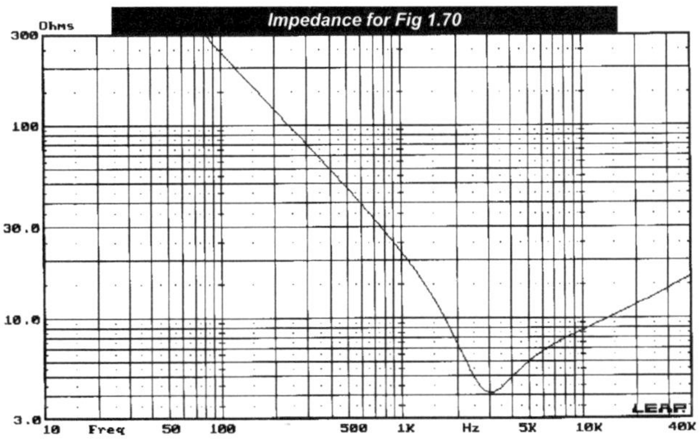

—— Tw #2 3rd 2 with cap p7

FIGURE 1.71

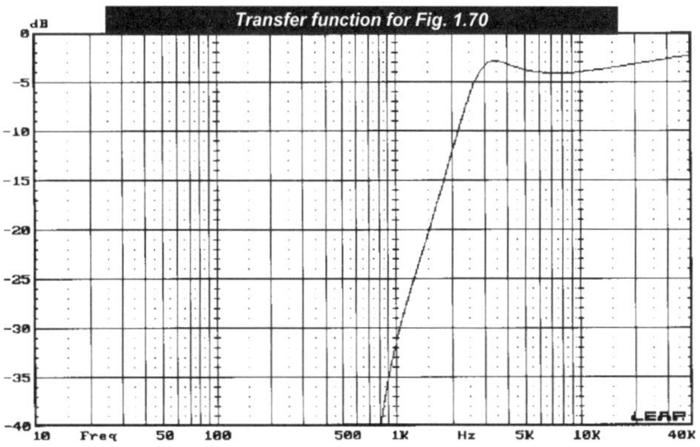

—— Tw #2 3rd 2 with cap p7

FIGURE 1.72

Tweeter circuit #5 using third-order high-pass + series attenuation resistor and conjugate capacitor

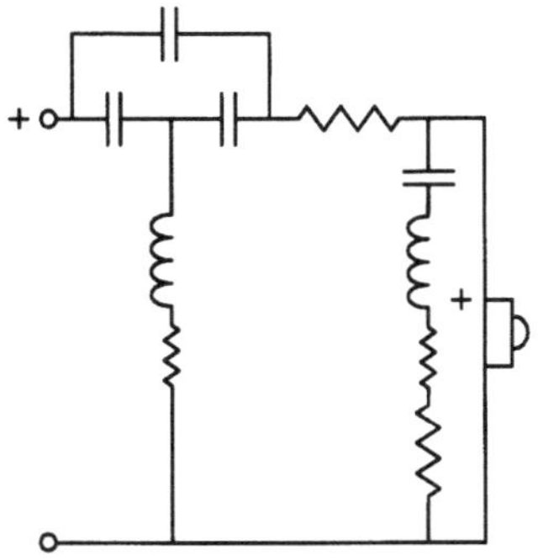

FIGURE 1.73

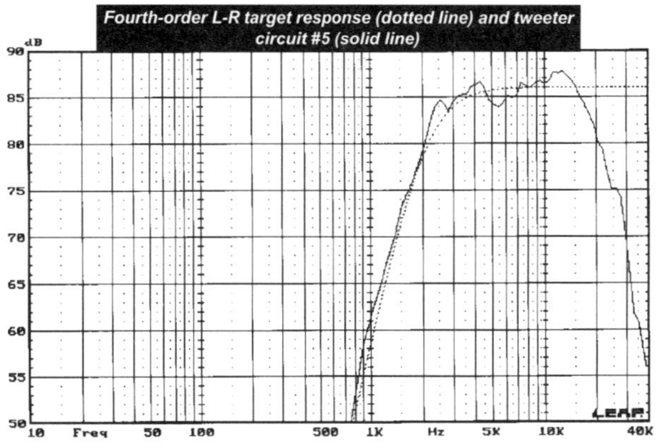

—— Tw 3rd + conj p8
.... Target

FIGURE 1.74

I have evaluated each of the circuits separately:

1. The second-order circuit with conjugate failed to provide sufficient attenuation in the network stopband. I could not optimize it closer than shown. The impedance drops to nearly 3Ω and the transfer function peaks to 1.5dB at the knee, indicating a relatively high Q.

2. The third-order topography incorporating an L-pad, which tends to swamp the driver resonance, closely follows the target. It had no transfer function peaks, but the impedance dropped to a low of 3Ω.

3. The third-order circuit, with a simple attenuation on the driver side of the high-pass section, locked onto the target to about -15dB. It

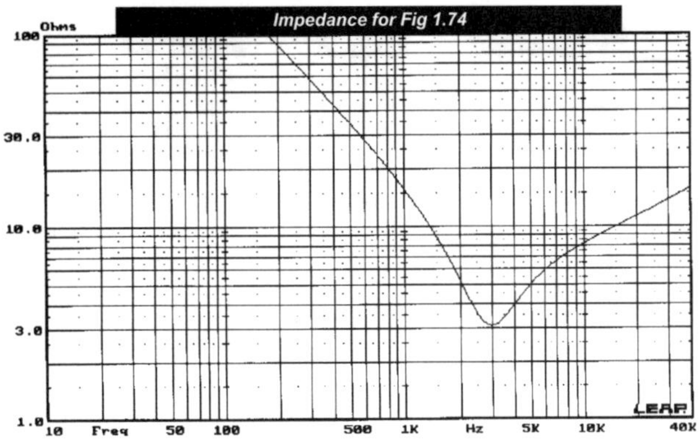

— Tw 3rd + conj p8

FIGURE 1.75

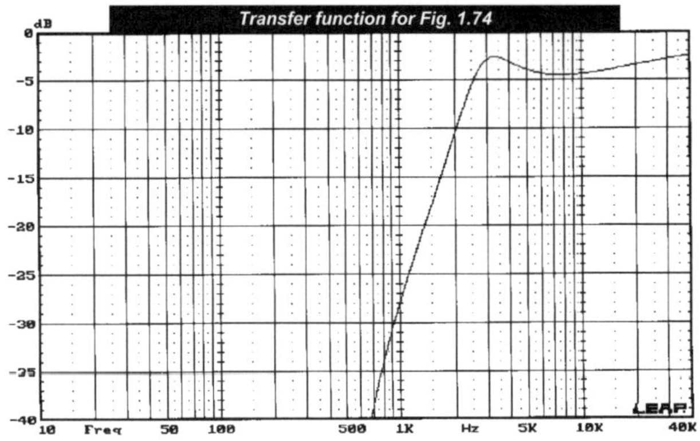

— Tw 3rd + conj p8

FIGURE 1.76

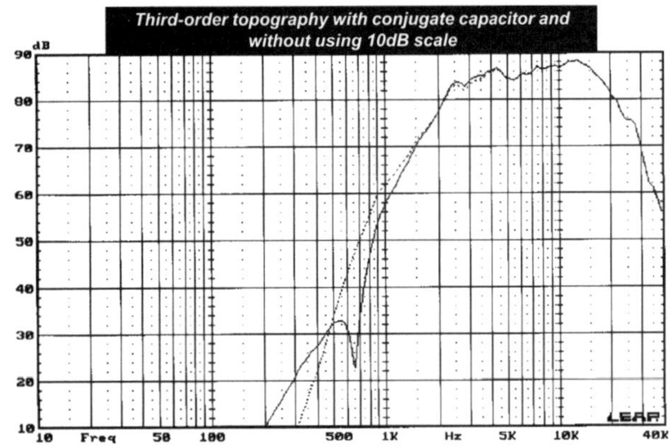

— Tw #2 3rd 2 with cap p7
······· Tw #2 3rd 2 no cap p5

FIGURE 1.77

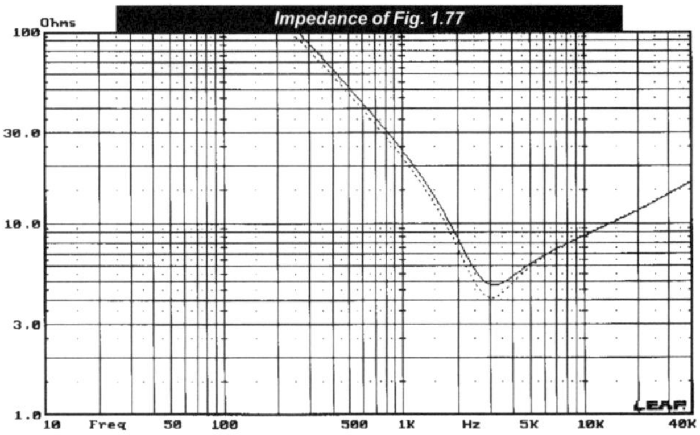

— Tw #2 3rd 2 no cap p5
······· Tw #2 3rd 2 with cap p7

FIGURE 1.78

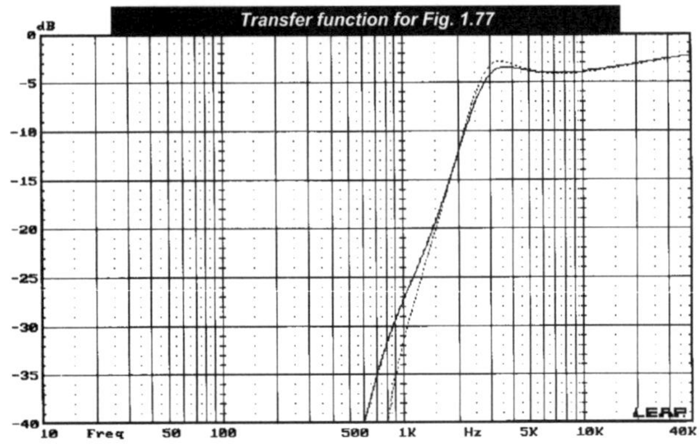

— Tw #2 3rd 2 no cap p5
······· Tw #2 3rd 2 with cap p7

FIGURE 1.79

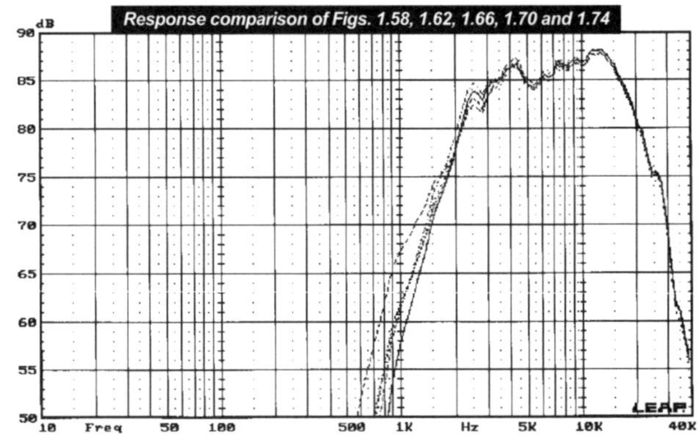

— Tw #2 3rd 2 with cap p7
······· Tw #2 3rd 2 no cap p5 -·-· Tw 3rd + conj p8
--- Tw #1 2nd + conj p4 - - - Tw 3rd + 1pad p9

FIGURE 1.80

then began to deviate. Low point for the circuit impedance was a safe 4.8Ω, and the transfer function had no pronounced peaking. The response is shown without the CR circuit switched in.

4. This circuit uses one capacitor shorting across two others in the third-order high-pass, a design similar to the KEF crossover.[2] The effect is similar to using an LCR conjugate tuned to the driver resonance. As depicted by the comparison between circuits 3 and 4 (*Figs. 1.77-1.79*), it results in better tracking of the target at lower frequencies.

Figure 1.77 has a 10dB rather than a 5dB scale, so it is easy to see the increased attenuation in the second octave below the crossover frequency. The

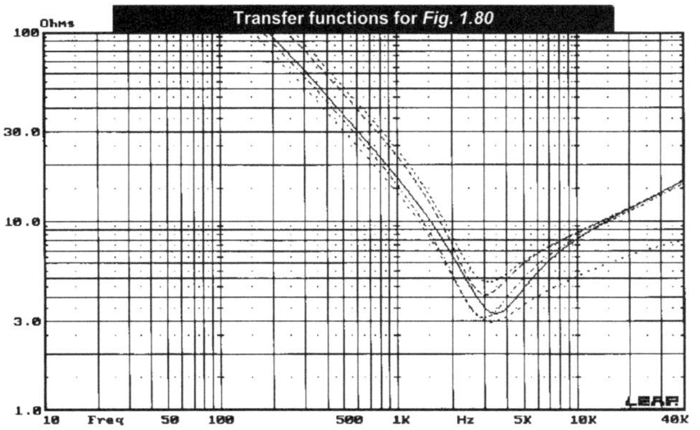

— Tw #1 2nd + conj p4
······· Tw #2 3rd 2 no cap p5
– – – Tw #2 3rd 2 with cap p7
–·–· Tw 3rd + conj p8
· · · Tw 3rd + 1pad p9

FIGURE 1.81

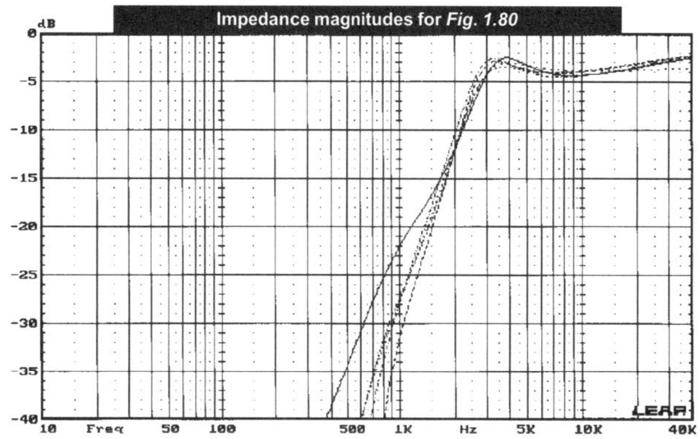

— Tw #1 2nd + conj p4
······· Tw #2 3rd 2 no cap p5
– – – Tw #2 3rd 2 with cap p7
–·–· Tw 3rd + conj p8
· · · Tw 3rd + 1pad p9

FIGURE 1.82

Final high-pass tweeter schematic with added boost circuit (switchable).

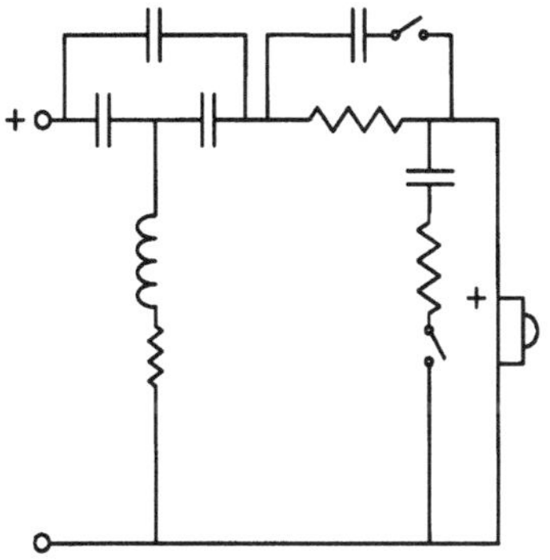

FIGURE 1.83

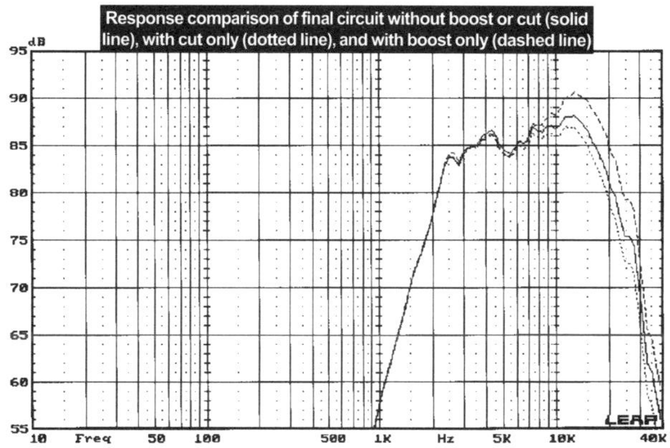

— Tw #2 3rd 2 with cap
······· Tw #2 3rd 2 with cap conj 1
– – – Tw #2 3rd 2 with cap conj 2

FIGURE 1.84

only penalty is a lower impedance minimum (4Ω) and increased peaking (1dB due to a higher Q) in the transfer function. The tradeoff is a less damped high-pass section for increased attenuation in the lower octaves. I chose this circuit, and opted for the increased attenuation.

Other alternatives, include splitting the attenuation resistor, and putting about 1Ω after the first (left side) capacitor. While this would not affect the response in the stopband, it would increase SPL by about 1dB between 2.8-3.5kHz, decrease the knee in the network transfer function by 1dB, and bump the impedance minimum back to 5Ω.

5. This circuit demonstrates the utility of computer optimization in testing unusual topographies. The third capacitor in the high-pass section is redundant when used with the LCR circuit. I tried relocating the LCR to further contour the circuit, but got extensive peaking in the transfer function and a low-impedance minimum of 3Ω.

Figures 1.80-1.82 compare the frequency response, impedance, and transfer functions of all five circuits. I finally chose the one shown in Fig. 1.83, which is a variation of circuit 4, with circuit 3's switchable CR conjugate added. As the various curves indicate, the tweeter's response increased 1-2dB in the 10kHz region, regardless of the network. The conjugate CR contours that response to the 86dB level, while the switch allows the option of disabling the circuit (sometimes thought to hamper upper harmonic detail). The capacitor parallel to the series resistor has the opposite effect—it provides a boost at the same frequencies.

As Fig. 1.84 illustrates, you can easily create a three-position contour of the tweeter's upper-frequency response with an on-off-on double-pole double-throw (DPDT) switch. Figures 1.85 and

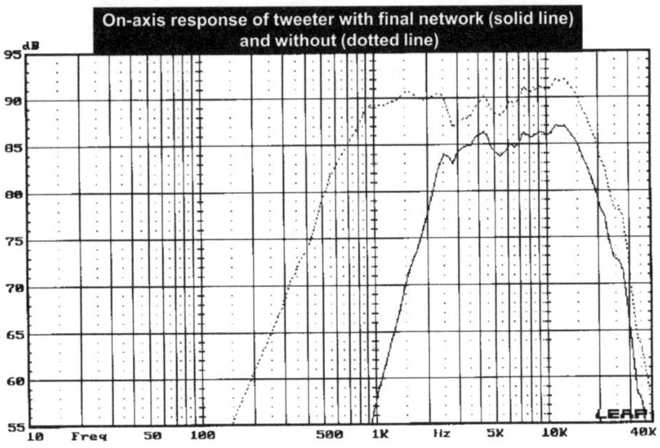

—— Tw #2 3rd 2 with cap conj 1
········ New dgl 14

FIGURE 1.85

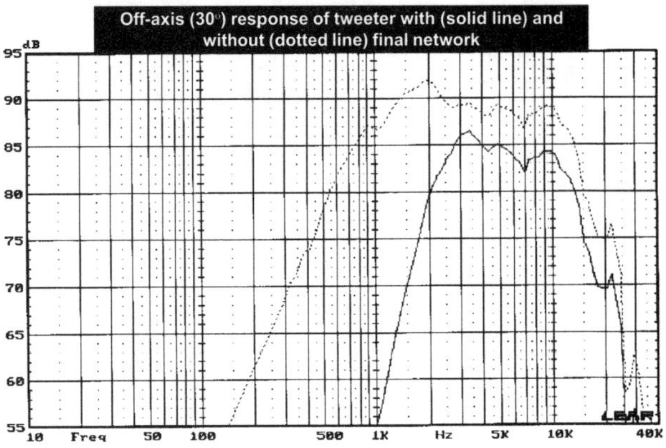

—— Tw #2 3rd 2 with cap conj 1
········ New dgl 14

FIGURE 1.86

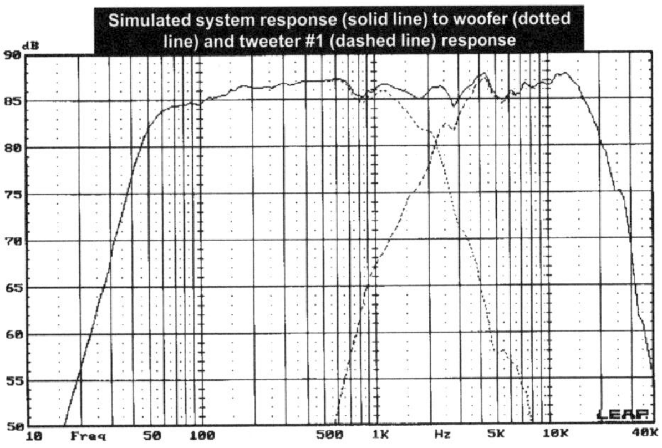

—— 8 + 11
········ Wof 2b 4" port p3
- - - Tw #1 2nd + conj p4

FIGURE 1.87

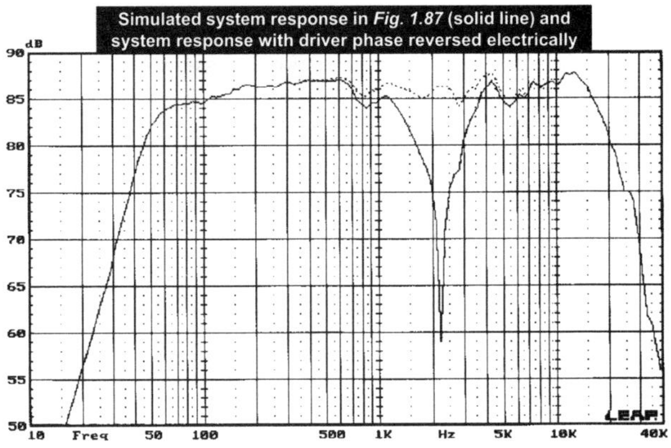

—— 8 + 11 rev
········ 8 + 11

FIGURE 1.88

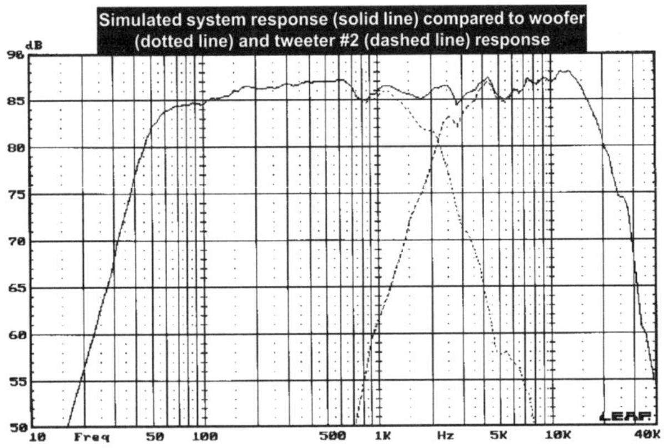

—— 8 + 16
········ Wof 2b 4" port p3
- - - Tw 3rd + 1pad p9

FIGURE 1.89

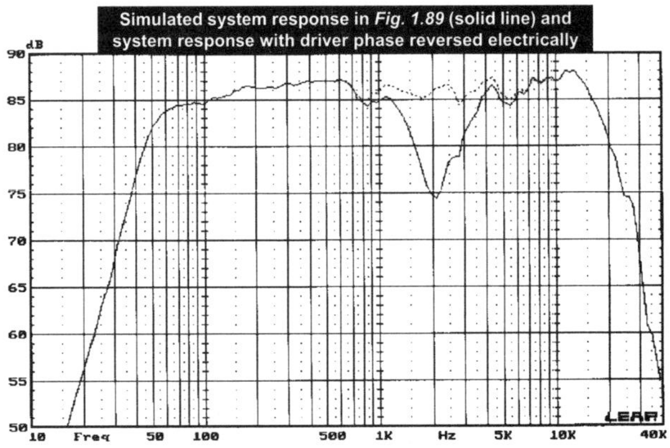

—— 8 + 16 rev
········ 8 + 16

FIGURE 1.90

1.86 compare the tweeter response with and without the high-pass network—both on and 30° off-axis.

1.70 System Analysis

I selected tweeter high-pass circuit 4, but looking at the summation of the various tweeter circuits with the woofer low-pass circuit is a good exercise.

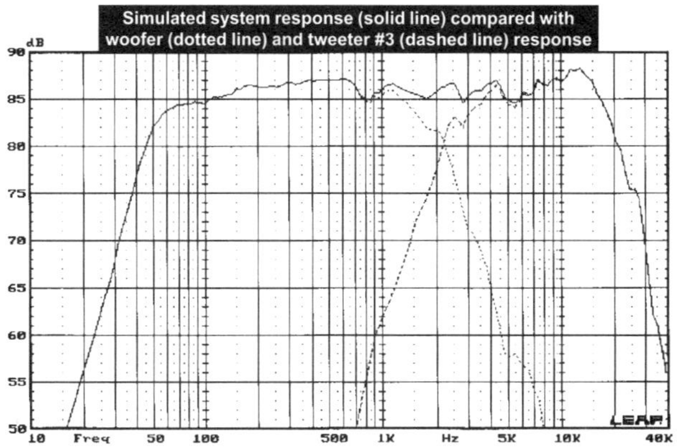

—— 8 + 13
.... Wof 2b 4" port p3
- - - Tw #2 3rd 2 no cap p5

FIGURE 1.91

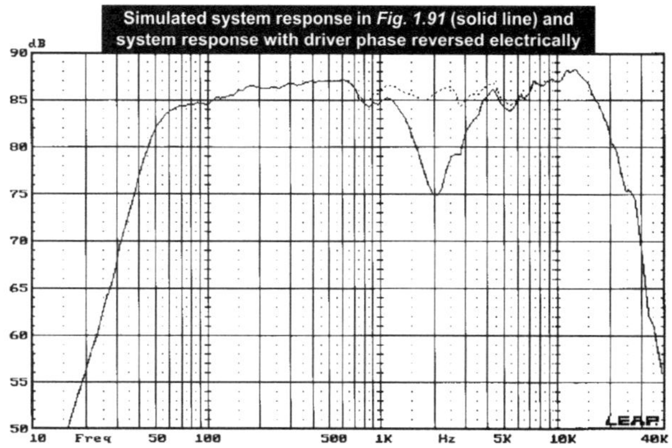

—— 8 + 13 rev
.... 8 + 13

FIGURE 1.92

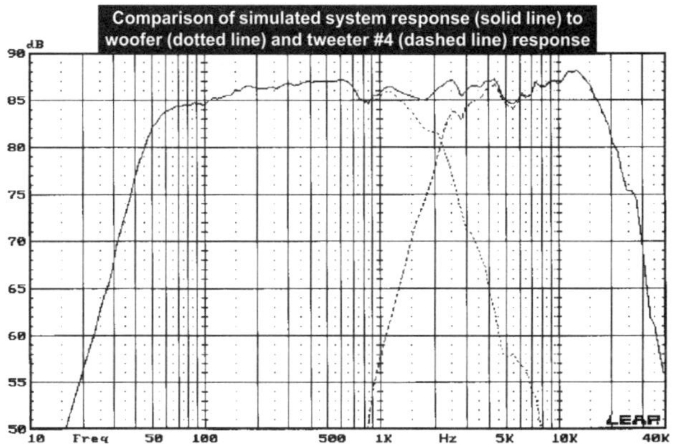

—— 8 + 14
.... Wof 2b 4" port p3
- - - Tw #2 3rd 2 with cap p7

FIGURE 1.93

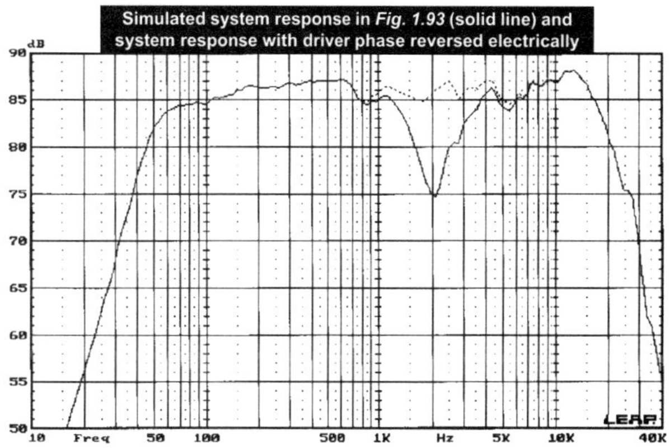

—— 8 + 14 rev
.... 8 + 14

FIGURE 1.94

As using all three woofer circuits would produce an excessive number of permutations, I restricted the comparison to combinations of the five tweeters and the #3 woofer's low-pass circuit.

A good way to evaluate the summation of high- and low-pass sections is to examine both the in-phase and out-of-phase summations (with the tweeter polarity reversed). The depth and symmetry of the null that results with the drivers out-of-phase indicates the degree and quality to which the two devices are operating in-phase.

The five combinations are given by these curves:

Tweeter #	In-phase	Out-of-phase
1	1.87	1.88
2	1.89	1.90
3	1.91	1.92
4	1.93	1.94
5	1.95	1.96

As the various curves show, each circuit combination produces a reasonably flat response. (The comparison of all combinations is shown in Fig. 1.97.) For this reason you must consider other criteria, such as nulls. Combination 1, for example, has a very deep null (closely in-phase) but not so symmetrical as those produced by combinations 3 and 4. The null in combination 5 is also deep but not as symmetrical as the others. While not a stringent criterion, this is an indication that the blend from high- to low-pass is providing an even transition.

The completed design uses woofer simulation #3 and tweeter simulation #4. Figure 1.98 is the full crossover schematic, while the values appear in Table 1.1. Figures 1.99-1.101 show the LEAP computer simulations of the completed speaker's on-axis, and 30° off-axis responses, and impedance.

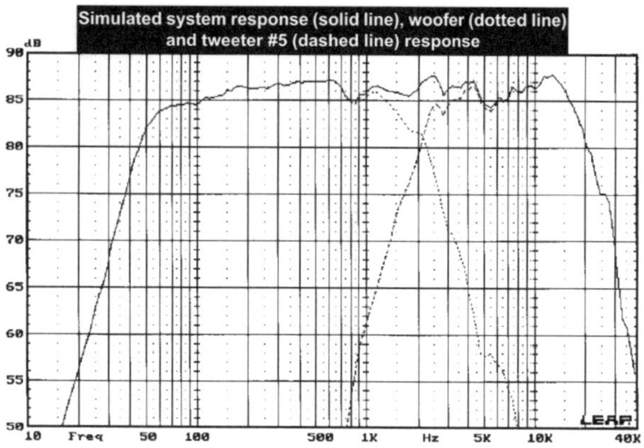

— 8 + 15
······· Wof 2b 4" port p3
- - - Tw 3rd + conj p8

FIGURE 1.95

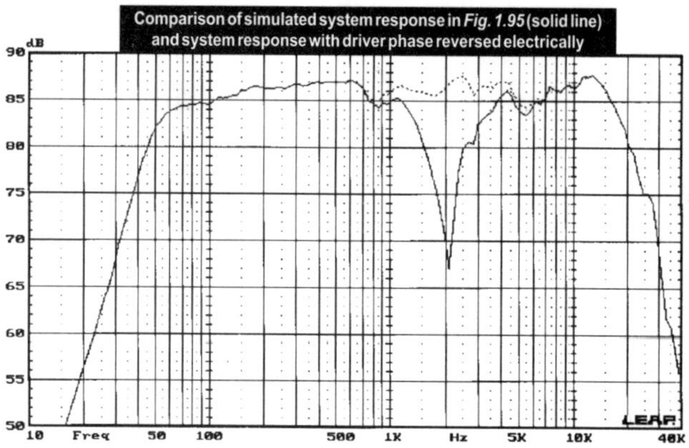

— 8 + 15 rev
······· 8 + 15

FIGURE 1.96

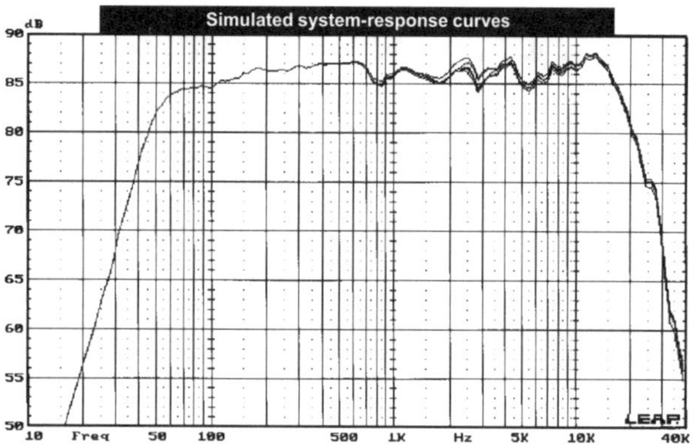

— 8 + 11
······· 8 + 13
- - - 8 + 14
-·-·- 8 + 15
·-·- 8 + 16

FIGURE 1.97

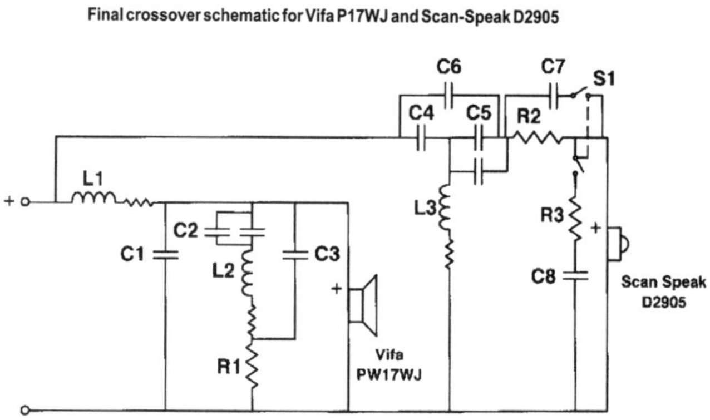

FIGURE 1.98

TABLE 1.1
Crossover Component Values

Low-pass section

	Value	Source
C1	20µF, ASC poly	NCMS
C2	47µF, NP electrolytic	IT (#1341075)
	1µF Sprague poly	NCMS
C3	4.7µF, Sprague poly	NCMS
L1	1.2mH, 0.16Ω	NCMS
L2	0.68mH, 0.41Ω	IT (#0022)
R1	10Ω, 10W, Ohmite	NCMS

High-pass section

C4	6µF, Sprague poly	NCMS
C5	8µF, 3.9µF Sprague	NCMS
C6	0.56µF poly	IT (#1341270)
C7	2µF Sprague poly	NCMS
C8	1µF Sprague poly	NCMS
L3	0.35mH, 0.08Ω	NCMS
R2	6.19Ω, 10W, Ohmite	NCMS
R3	6.19Ω, 10W, Ohmite	NCMS
S1	on-off-on DPDT sw.	

NCMS = North Creek Music Systems
IT = InterTechnik
Poly = Polypropylene

Note: When I built the prototype crossovers, I did not realize North Creek made smaller-gauge wire inductors. I could have substituted a higher-DCR inductor for L3 with minimal consequence. Also, 5W is generally considered sufficient for tweeter attenuation resistors, although the Ohmite resistors, which come only in 10W values, are noninductive and quite good.

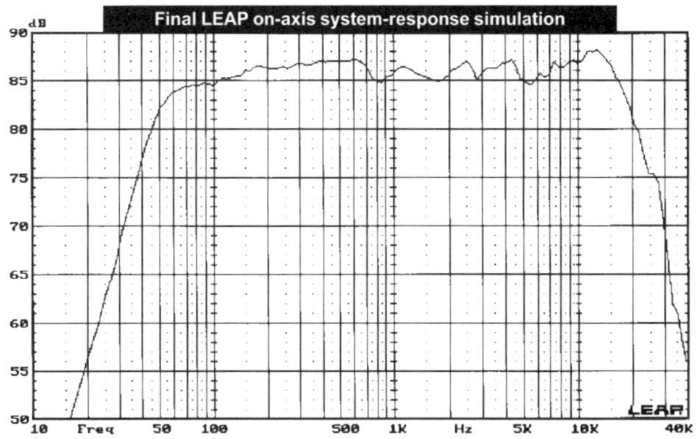

—— Wof 2b 4" port p3
.... Tw #2 3rd 2 with cap p7

FIGURE 1.99

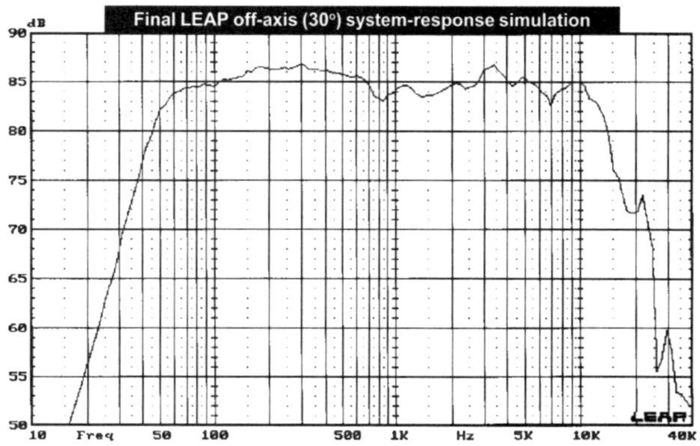

—— Wof 2b 4" port p3
.... Tw #2 3rd 2 with cap p7

FIGURE 1.100

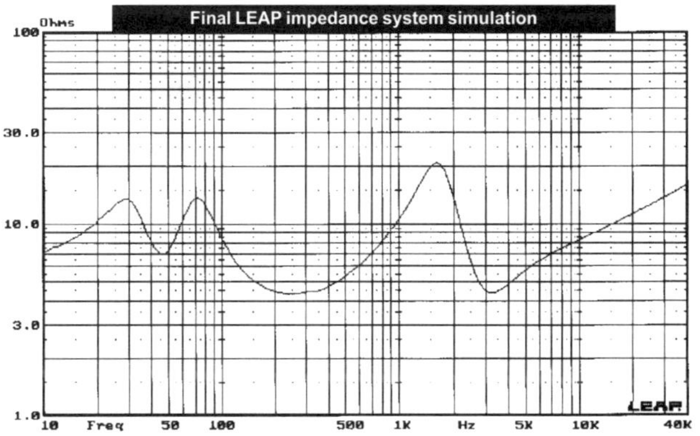

—— Wof 2b 4" port p3
.... Tw #2 3rd 2 with cap p7

FIGURE 1.101

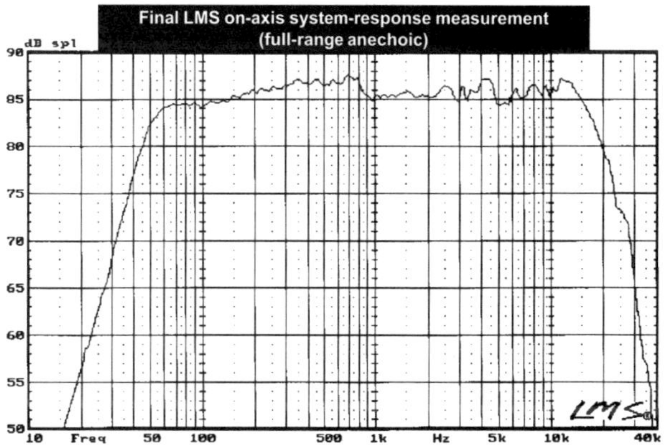

- - - FR with conj

FIGURE 1.102

1.80 Measured Performance

Does the constructed prototype measure close to the computer simulations? More importantly, does it produce acceptable sound without the need for alteration? The answer to both questions is "yes."

Figure 1.102 illustrates the prototype's on-axis frequency response. With the speaker tilted 6°, the response is ±1.6dB at 66Hz-16kHz and ±3dB at 48-18.8kHz—certainly respectable for a two-way 6.5" speaker. Furthermore, it deviates only slightly from the simulation shown in Fig. 1.99.

The simulations might have come even closer had I modeled the capacitor-loss variables. Thanks to the low loss nature of high-quality capacitors, however, and the relative simplicity of two-way crossovers, I was able to skip that process. With more complicated circuits, computer network optimization requires increased attention to component losses and nonlinearities. Bandpass and lower-frequency networks, for instance, may require some types of cored inductors and other nonlinear components. In addition, many of the capacitor values in this book were made from multiple paralleled capacitors, which substantially decreases the effects of loss or induction in individual capacitor elements.

The off-axis response (Fig. 1.103) tells an even better story. It remains ±2dB throughout most of its range—even at 30° off-axis. From 0°-45° the picture begins to deteriorate, but still maintains ±2.5dB to 10kHz (Fig. 1.104). The 0-90° plot in Fig. 1.105 gives a more complete picture of this speaker's horizontal dispersion. Maintaining a consistent response in the horizontal plane to 45° is desirable, as it significantly increases the smoothness of the homogenized blend of direct and reflected sound that reaches the listener's ear in the typical room (if such a thing exists).

In the vertical plane, the speaker exhibits the lobing typical of noncoincident driver mounting. Figure 1.106 shows the vertical polar frequency response to 30° going over the top from an on-axis position. Figure 1.107 is the same picture going under the speaker (for the sake of simplicity, vertical

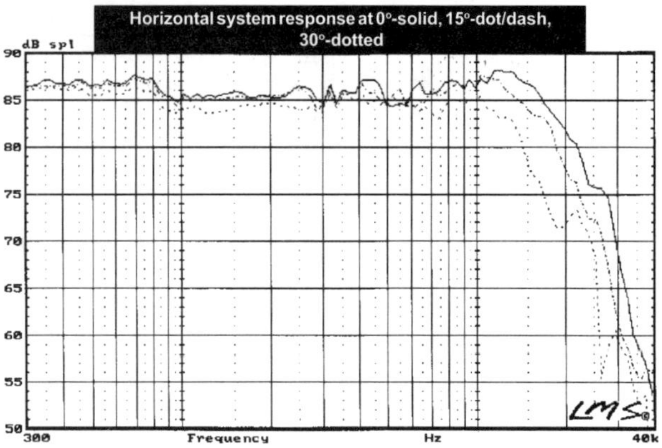

—— Mla 0 v
-·-· Mla 15
- - - Mla 30

FIGURE 1.103

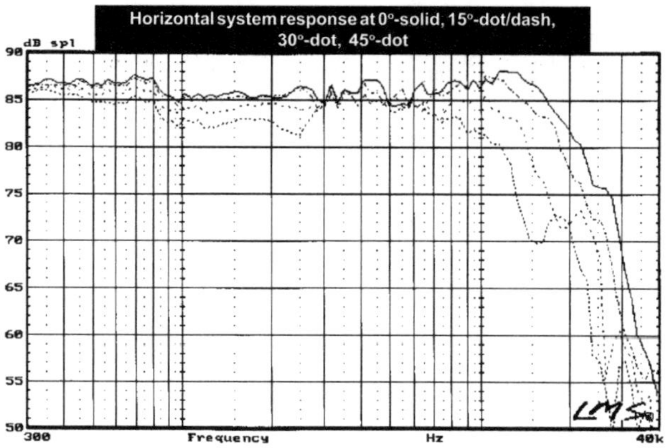

—— Mla 0 v
········ Mla 45
-·-· Mla 15
- - - Mla 30

FIGURE 1.104

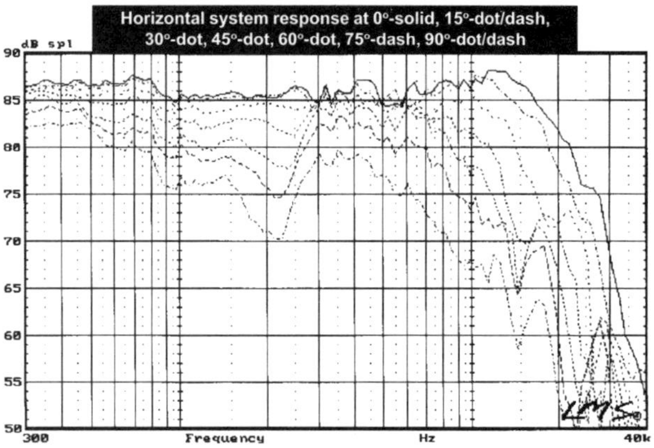

—— Mla 0 v ········ Mla 60
········ Mla 45 --- Mla 75
-·-· Mla 15 -·-· Mla 90
- - - Mla 30

FIGURE 1.105

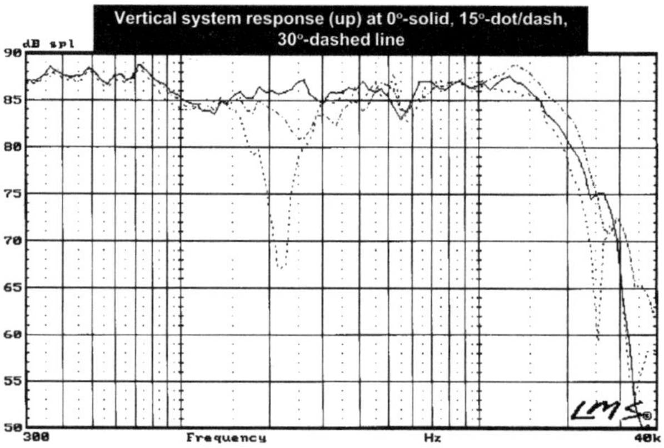

—— Mla 0 H
-·-· Mla + 15
- - - Mla + 30

FIGURE 1.106

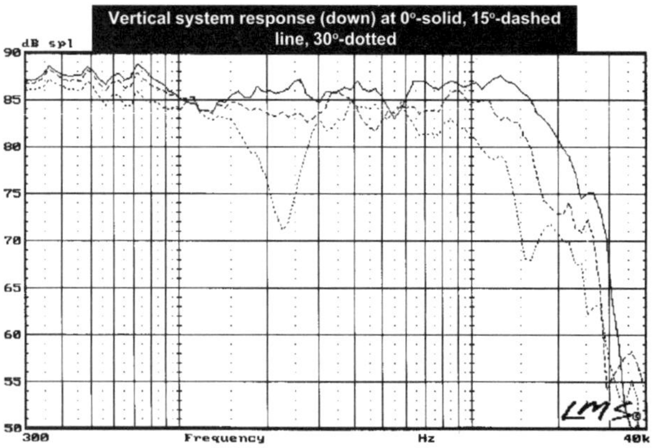

—— Mla 0 H
--- Mla -15
········ Mla -30

FIGURE 1.107A

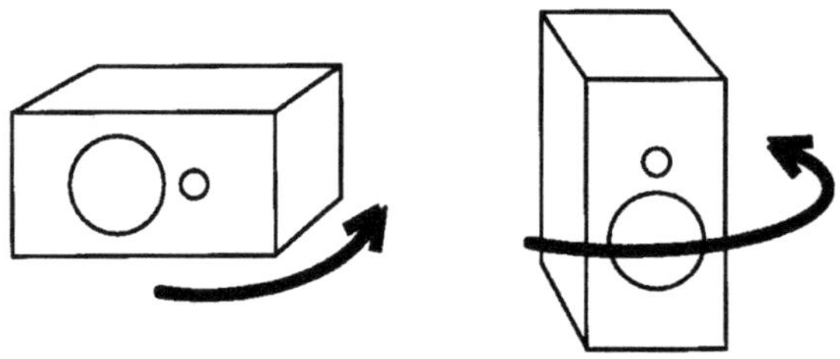

FIGURE 1.107B

polar measurements are made by changing the speaker's orientation and placing it horizontally on a revolving platter rather than by moving the microphone in the vertical plane).

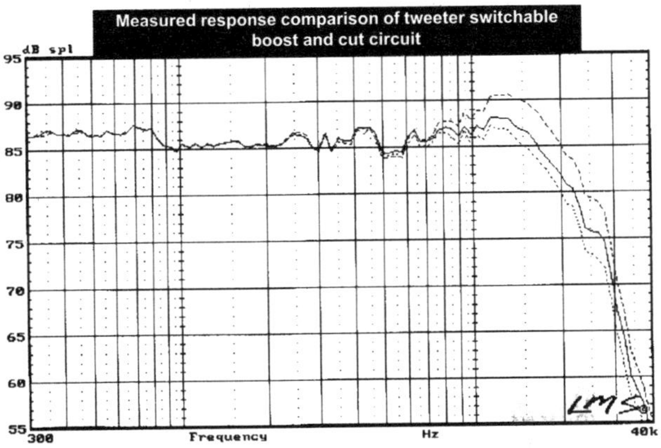

— Mla 0 V
······· Mla 0 v tw conj1
- - - Mla 0 v tw conj2

FIGURE 1.108

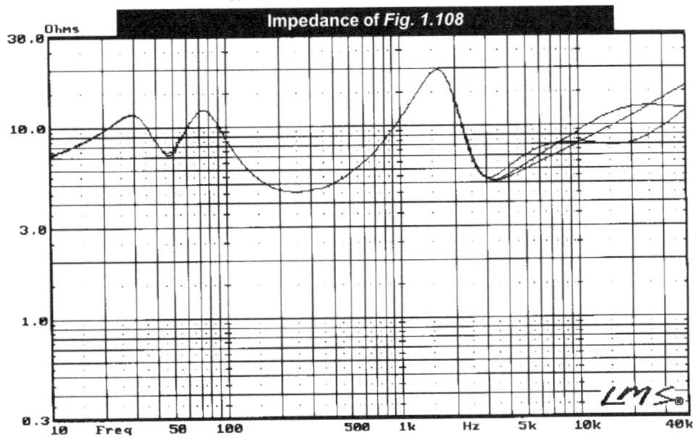

— M1 sys imp
· · · M1 sys imp tw -
······· M1 sys imp tw +

FIGURE 1.109

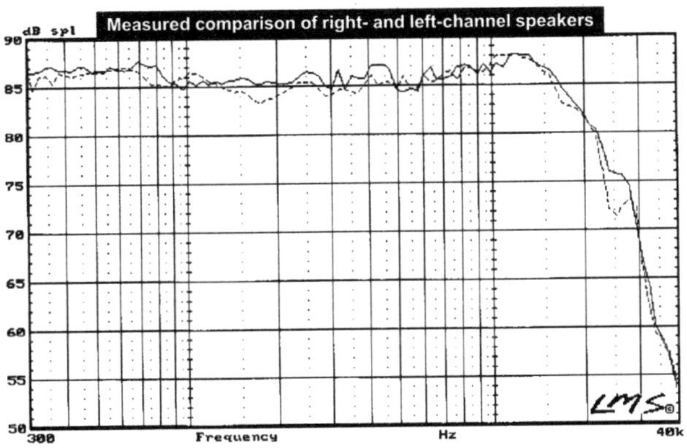

— M1b conj
- - - M1a 0 v

FIGURE 1.110

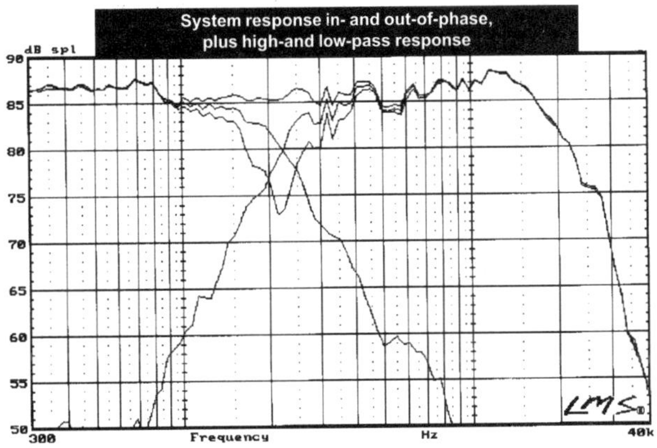

— M1a 0 v
— M1a 0 wof
· · · M1a 0 tw
······· M1a rev

FIGURE 1.111

The tweeter contour switching works as expected, as seen in the response changes for different switch positions (Fig. 1.108) and the associated impedance changes (Fig. 1.109). Comparison of the two completed speakers is displayed in Fig. 1.110. Deviation is no greater than 1.5dB at 1.7kHz, which is not bad for unsorted, unmatched, off-the-shelf drivers. Your experience with this design should be similar, barring changes in the manufacturers' production from batch to batch.

Figure 1.111 gives the in- and out-of-phase responses of the speaker, plus the individual driver and network responses. It looks nearly identical to the simulation and shows the even, symmetrical slopes and out-of-phase null, which occurs at the 2.2kHz crossover frequency. I calculated the speaker's acoustic phase (Fig. 1.112) using the technique described for determining time delay between drivers.

The last two performance graphs are the complex impedance curve and the MLSSA cumulative spectral-decay curve (Fig. 1.113, Fig. 1.114, respectively). The former gives a maximum inductive phase angle of 36.56° at 867Hz, and maximum capacitive phase angles of 46.13° at 2302Hz and 35.05° at 99.6Hz. This performance should not cause a problem with any amplifier driving the speaker, although the minimum impedance of 4.5Ω at 250Hz precludes paralleling two of these speakers. The range of impedance is from 20Ω to 4.5Ω, a ratio of 4.44:1, which means the voltage divider effect would cause imperceptible response changes with 14-gauge or greater speaker cable. The MLSSA time-decay plot in Fig. 1.114 shows no major resonances, and all midrange activity is below 1.3μs and uncluttered in appearance.

1.90 Cabinet Performance

A speaker's construction is nearly as important in determining its overall sound quality as the rest of the design work. The goal of cabinet design is to limit radiation in the room to that produced by the

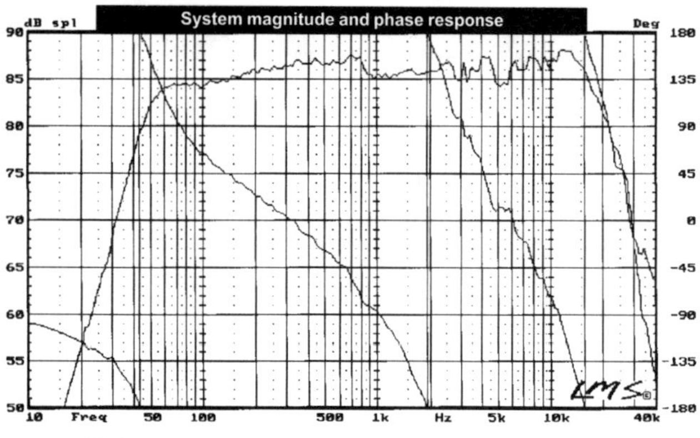

— Fr phase

FIGURE 1.112

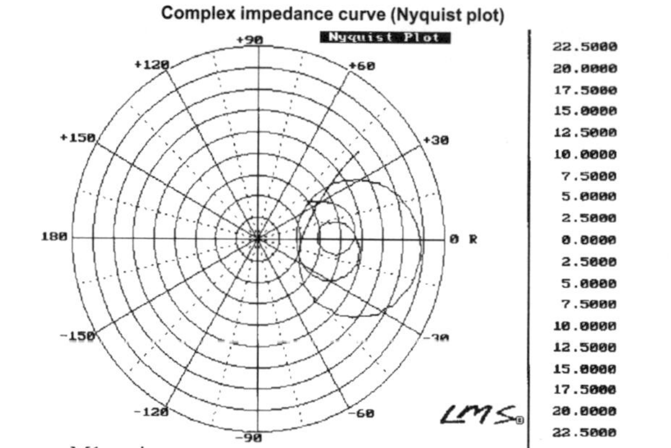

— M1 sys imp

FIGURE 1.113

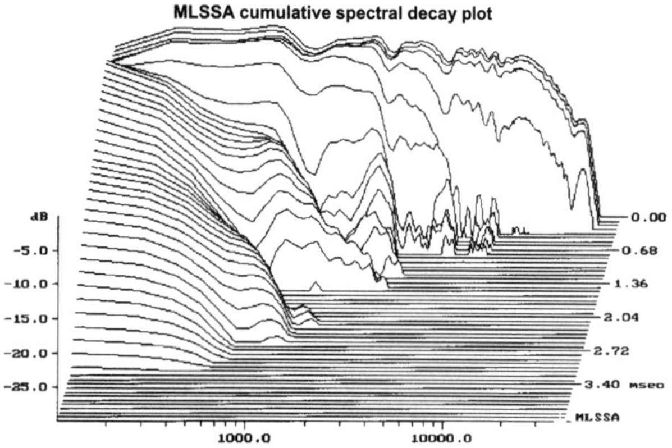

FIGURE 1.114

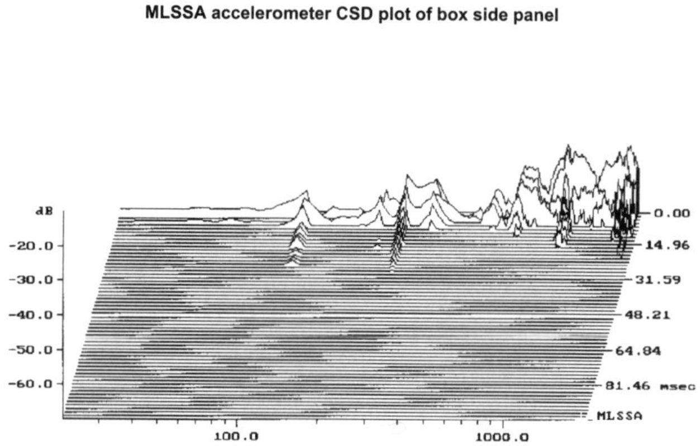

FIGURE 1.115

drivers, and to limit cabinet diffraction effects on frequency response. For all four speakers in this book I designed cabinets which minimize diffraction and are as free as possible from extraneous vibration.

The effects of cabinet diffraction on frequency response are illustrated by the various frequency-response curves. The only step I took to reduce diffraction was to place the tweeter off-center on the baffle. (One cabinet had the tweeter mounted to the left and the other to the right for "mirror imaging.") The enclosure's rounded corners are mostly cosmetic, primarily because the oak quarter round's small radius has minimal, if any, effect on diffraction. The other obvious diffraction problem is the woofers' rising response, which I compensated for in the crossover design.

I measured the effectiveness of the various cabinet damping techniques with an accelerometer. Figures 1.115-1.117 give the cumulative spectral-decay (CSD) plots (using the MLSSA FFT analyzer) with the accelerometer mounted on the side, top, and rear centers of the enclosure, which was placed on the stand. The piezo-element accelerometer, described in the Fourth Edition of The Loudspeaker Design Cookbook, is highly effective but uncalibrated. (Its low-frequency performance is superior to the polyvinylidene-fluoride accelerometer described in Voice Coil, November 1991.) Since it is uncalibrated, the piezo accelerometer can be used only for relative information.

To illustrate the effectiveness of the cabinet construction described below, Figs. 1.118-1.120 show the same accelerometer curves for a mid-price commercial loudspeaker. (At $500/pair, the commercial speaker may be slightly less expensive than building any of the speakers in this book, depending on your skill and resourcefulness.) The commercial speaker is a typical production unit built with 0.75" medium-density fiberboard (MDF) and with no bracing or additional damping material. Thanks to its battleship-style construction, the P17WJ/D2905 prototype's accelerometer curves look considerably better than those of the off-the-shelf unit. The decrease in midrange coloration

MLSSA accelerometer CSD plot of box top panel

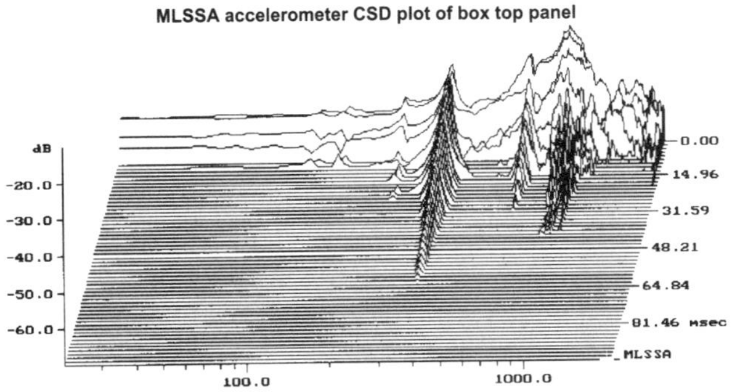

FIGURE 1.116

MLSSA accelerometer CSD plot of box rear panel

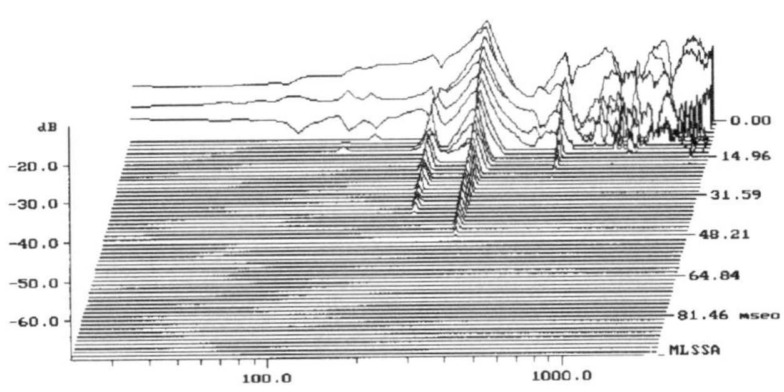

FIGURE 1.117

MLSSA acceleromerer CSD plot of box side panel for typical production loudspeaker

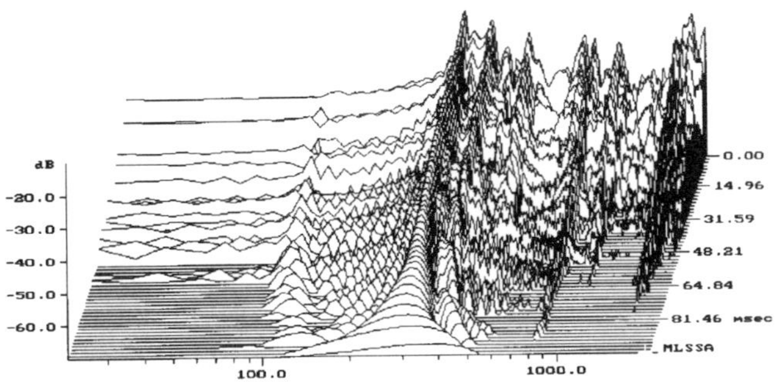

FIGURE 1.118

MLSSA accelerometer CSD plot of box top panel for typical production loudspeaker

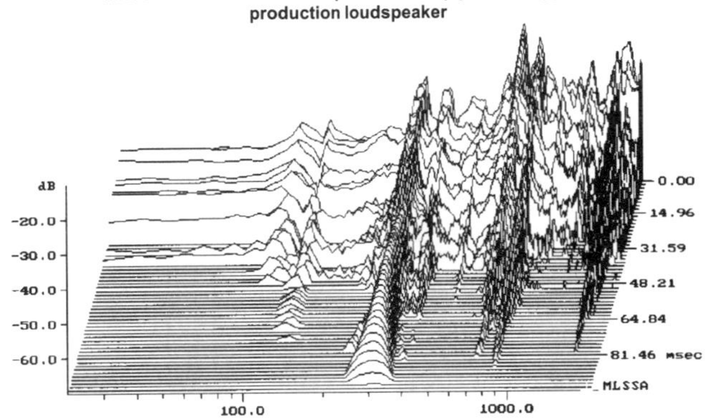

FIGURE 1.119

MLSSA accelomerer CSD plot of box rear panel for typical production loudspeaker

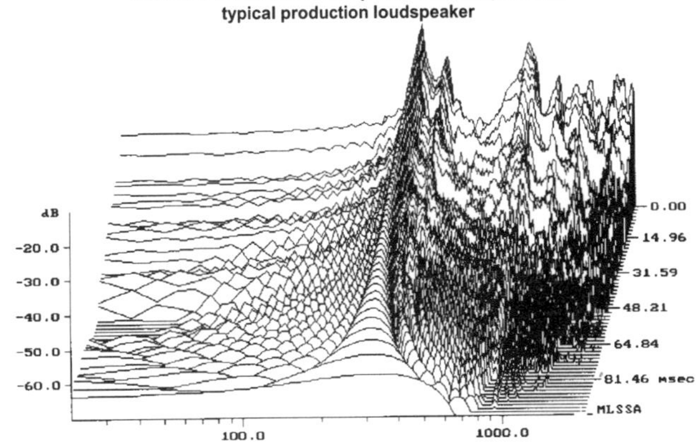

FIGURE 1.120

using the type of cabinet construction I describe is substantial.

I made a more realistic comparison using LMS. The MLSSA curves are of acceleration with frequency, and will cause upper frequencies to appear higher in amplitude than lower frequencies; however, the opposite is true. Lower frequencies cause the box panel to move physically farther than do higher frequencies. To alter these curves to reflect excursion, divide the accelerometer curves by frequency twice. Figure 1.121 compares the altered LMS accelerometer curves of the prototype and production speakers which shows the low frequencies to be greater in amplitude than the high frequencies and more in line with the panel's actual excursion. Note also the prototype's decreased amplitude throughout most of the test range.

1.100 Construction Details

While this book's primary purpose is to explore loudspeaker design methodology, I have provided sufficient construction detail should you wish to duplicate the designs.

Cabinet construction for the first speaker is diagrammed in Fig. 1.122. While its damping characteristics are quite good, this cabinet design uses relatively simple construction techniques. I cut the top, bottom, rear baffle, and left and right walls from 1" MDF, and assembled them using glue and staples (nails are also appropriate). Rather than mitering the four walls, I used butt-joints for easier assembly.

To limit internal reflections and standing wave modes, I used an internal rear enclosure wall

LMS accelerometer curves divided by frequency twice comparing side panel test of a typical production loudspeaker-dotted and the P17WJ/D2905 prototype speaker (solid line)

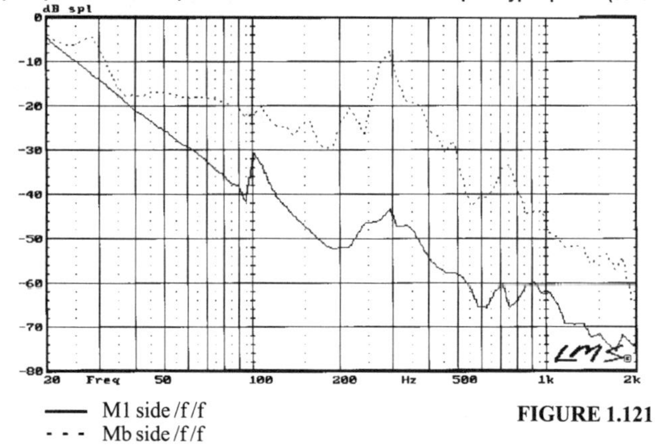

FIGURE 1.121

Box construction details

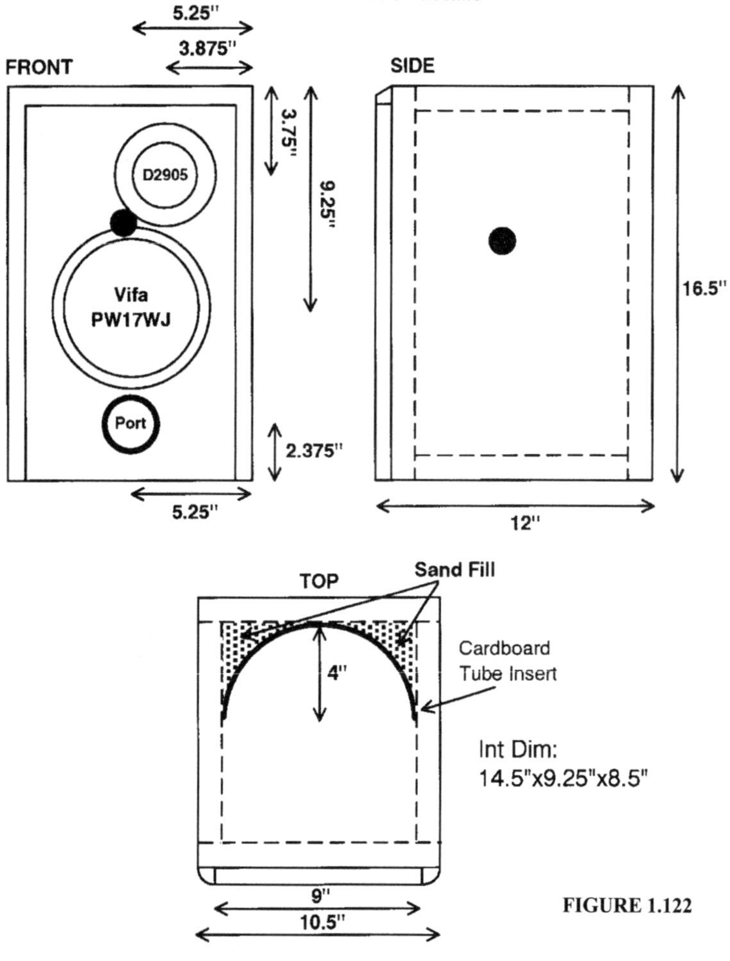

FIGURE 1.122

made from 8-inch-diameter cardboard tubing (Sonotube). I cut the 0.25-inch-thick cardboard in half lengthwise to form a semicylinder, and trimmed it to the inside height of the box. I positioned it at the rear of the enclosure and stapled it to the back and side walls. I then sealed the cardboard edges using a combination of hot glue and silicone, which formed two empty spaces in the corners of the box.

After assembling and finishing were completed, I drilled 0.5" holes in the bottom of the enclosure at the center of the corner spaces. I poured fine white sand (the kind used for gardening) through the holes into the cavities behind the cardboard tubing. Once these were completely filled, I sealed the holes with dowel plugs. If you use a terminal cup, it must be located to the side (not the center) of the enclosure's rear panel, so as to clear the cardboard semicylinder. (Incidentally, I made this internal semicylinder concept as a prototype several years ago—before the release of the Genesis speaker, which uses the same format, without sand backfill.)

Bracing consists of two 1.125" hardwood dowels. I centered one dowel between the woofer and tweeter, extending it through the cardboard tube from the front to the rear baffle. (I cut the hole before mounting the tube.) The second dowel brace is slightly below the first, from side wall to side wall. (Brace locations are noted by small black circles in the diagram.)

Cutting the dowels exactly to length is important, albeit tricky. Dowels which are too long cause the enclosure to bow outward, while too short causes the walls to cave inward, especially if they are nailed or stapled through the wall. If you drill a shallow 0.25" countersink into each wall to position the dowels, however, you can cut them that much longer than the enclosure's dimension. I applied construction adhesive to the dowel insets, then positioned the dowels during the box-assembly process. After the adhesive set, I tied the bracings securely to the enclosure.

The front baffle is 1.75" deep and was made from a bottom and top layer of 1" and 0.75" MDF respectively. I assembled the baffle by first gluing and stapling the 1" layer to the box, and the 0.75" to the 1" layer. I routed the top section to countersink the drivers flush with the baffle surface.

To prevent the woofer from acoustically loading within the 1.75" channel, I used a curved 45° bit to rout the 1" baffle section on the inside of the woofer hole opening into the enclosure. The 0.75" layer has a 0.75" inset to accommodate both the 0.75" oak quarter-round trim on the baffle's sides and the angled oak trim piece at the top. (Be certain to measure the dimension of the oak quarter-round carefully, as it can vary considerably.) The detail of the top trim piece, which extends across the top of the front baffle's 0.75" layer, is illustrated in Fig. 1.123. The quarter-round is cut at the same angle as the top trim piece for a smooth curve across the top of the enclosure front.

Once the cabinet was assembled, I painted the front baffle and enclosure bottom. With contact cement I attached flat black laminate to the sides, top, and rear, and then glued the oak trim in place. I further damped the internal wall panels by covering 50-60% of the sides, top, and bottom (excluding the cardboard tubing) with two layers of Antiphon Type A-13 extensional damping. This 1/16 inch-thick, black bituminous felt-and-clay material has self-adhesive backing. An alternate, perhaps more effective, material, not available when I was constructing the prototype, is Black Hole Pad. Sold by Orca Design & Manufacturing, it is so thick that a single layer is effective.

For damping of standing waves within the enclosure, I used AC Foam (an egg-crate type of acoustic foam provided by Audio Concepts Inc.) to line the four walls and rear of the enclosure. I did not use it on the area immediately behind the port as the foam should not restrict the air movement near the port. I cut the port tube, fabricated from 2" ID black ABS plastic water pipe, to a 4" length and mounted it as shown in Fig. 1.122.

To mount the woofer and tweeter, I used Well-Nut inserts, from Pop Fastener Corp. These rubber inserts have nuts mounted in the end and are frequently used as vibration-damping mounts for electric motors. Well-Nuts mount from the top of the hole and allow for easy removal of the driver; they also provide additional isolation for the speaker from the cabinet.

Along with the Well-Nuts I used a foam tape gasket (the type used for weather stripping). A better choice is the neoprene-rubber tape gasket available from North Creek Music Systems.

Unlike most speakers, the crossovers for this design are located within the stem portion of the stand rather than inside the enclosure. This isolates the components from vibration and the drivers' magnetic fields, and also allows for a much larger component layout which enables the builder to separate the inductors to safe noninteractive distances. Two terminal sets—one each for the woofer and the tweeter—are therefore required on the rear of the enclosure.

I used a round terminal cup, provided by ACI, which comes fitted with a pair of gold binding posts. Because the cup is small (it fits a 2-7/8" hole), it is less prone to vibration than some large plastic cups which act as passive radiators. I added a second pair of gold binding posts from Madisound, identical to those on the ACI cup. The resulting four terminals provide direct connection to both woofer and tweeter.

The wiring is routed from the terminal cup through a hole in the Sonotube which is sealed with silicone to prevent sand from spilling into the box. Even though this box is vented, laying a small bead of silicone along all internal joints is a good idea.

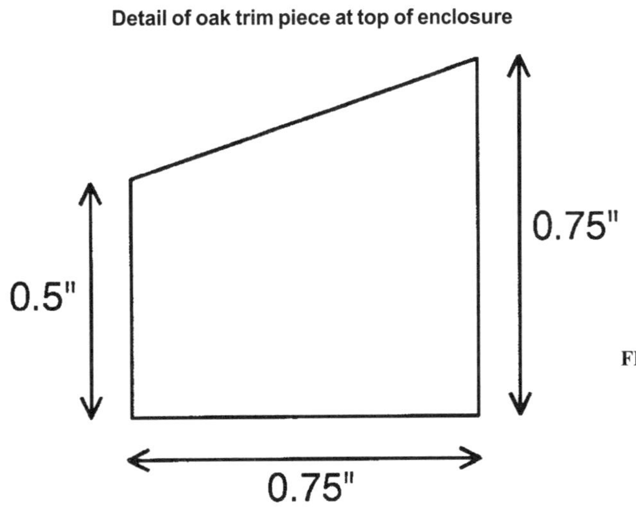

FIGURE 1.123

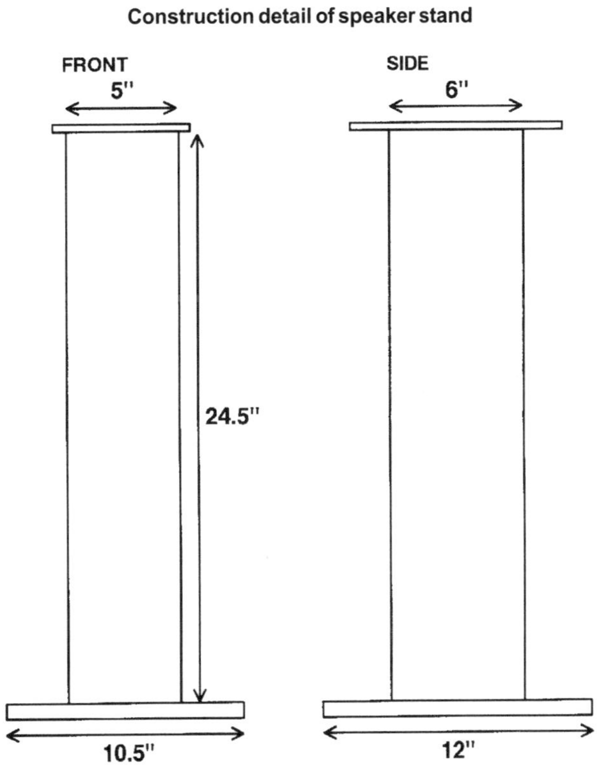

FIGURE 1.124

I fabricated grille frames, from 0.25" MDF, including standoffs (0.5"x0.5" pieces of the MDF glued to the frame). The standoffs lend an attractive look, and also provide a means to mount the snap-type plastic grille fasteners, from A&S Speakers. I attached black double-knit cloth to the frame using 3M's spray fabric adhesive. Although this method works well, you can easily alter it.

Construction detail of speaker stand top

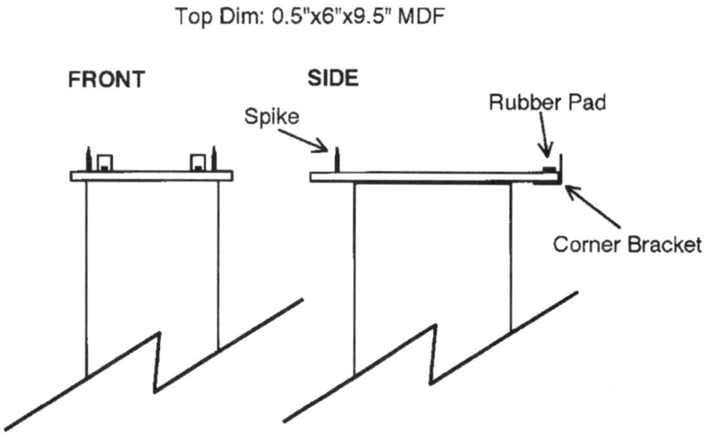

FIGURE 1.125

Photograph of crossover layout

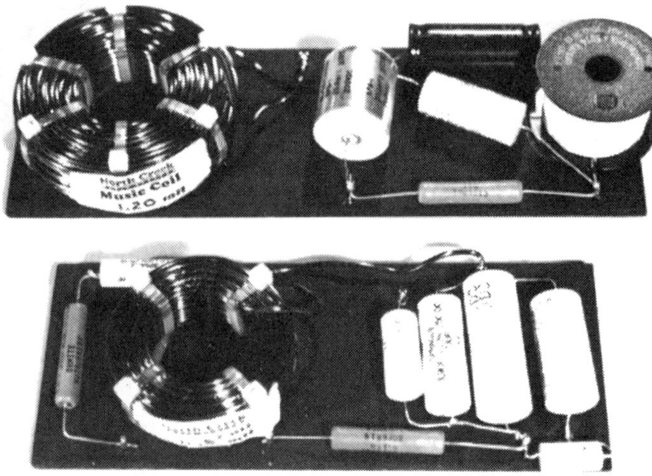

FIGURE 1.126

Photograph of TDM electronic two-way crossover

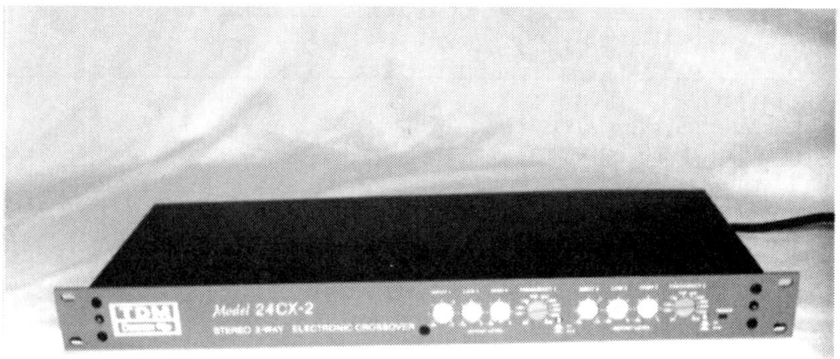

FIGURE 1.127

For the stands I used 0.75" MDF for both the base and the vertical stem (Fig. 1.124). I finished the base in black laminate and applied iron-on oak tape to the edges. To mount brass cone-type feet (also from A&S) to the bottom of the base, I drilled shallow countersink holes. I attached the base to the stem with 1.5" black wood screws, and finished it in black laminate to match the speakers. The inside walls of the stem are covered with Black Hole Pad to further reduce vibration transferred from the speaker cabinet.

I used a rectangular blank terminal cup to mount the three pairs of gold binding posts: one for the input, another for the crossover's high-pass section to connect the tweeter, and a third for the low-pass section to connect the woofer. To the cup panel I also mounted the tweeter's high-frequency contour switch. Obviously, another pair of terminals could easily be added to provide biwire capability. I used banana-plug jumpers to connect the crossover in the base to the drivers in the speaker enclosure.

The stand's top is detailed in Fig. 1.125. The screw-in type spike uses a brass insert, provided by Meniscus. You can adjust the spike until the speaker front reaches the required 6° tilt. Both the spike and the rubber pad help isolate the speaker from the stand; the corner bracket, installed at the rear of the top piece, prevents it from moving off the stand. The top piece is mounted to the stem and to the base in the same manner—with four wood screws. I added rope caulk (the kind used to seal windows and doors), as a gasket between the top and the vertical stem, to further isolate the speaker from the stand.

The crossover is mounted on 0.25" hardboard cut to fit the stem's interior side wall. I mounted high- and low-pass sections on separate boards, although I could have placed them on a single board. My intention was to use the stand's entire height to spread the inductors as far apart as possible so I did not take the normal precaution of placing adjacent ones at right angles and perpendicular to each other. (Even though they are mounted in the same plane, inductive coupling did not appear to be a problem when I measured individual sections.) By so mounting the inductors and securing them with some kind of strapping however, you can provide about 10-20dB more attenuation.

Layout of the other components is not critical (Fig. 1.126). Using silicone adhesive, I mounted the woofer section to the stem's interior side wall at the base of the stand. I affixed the tweeter section in the same manner with the inductor at the top of the stem. For internal wiring I used Monster cable for the woofer and Monster XP for the tweeter section.

I wired crossover components point-to-point. Brass split rivets provide connection points as well as a physical mount for capacitors and resistors. After laying out the components on the board, I located junction points, marked them with a pencil, drilled the board, and inserted the split rivets. I attached inductors with silicone adhesive.

1.110 Adding a Subwoofer

Many 6.5" drivers are incapable of providing the depth and dynamic impact of the lower bass octaves. Since the subject of this book is two-way speakers, it would be out of context for me to offer a subwoofer using a passive network. I do, however, provide a subwoofer using a commercial active network.

Subwoofer details are given in Chapter 3. (The speaker in that chapter, with its f_3 of 120Hz, was designed expressly for the subwoofer and cannot be used without it.) The crossover I used is the TDM Model 24CX-2, which provides 24dB/octave Linkwitz-Riley slopes with variable frequency controls (Fig. 1.127). This good-sounding electronic crossover uses biamping, which is a superior way to add a subwoofer to a speaker. While I have produced passive networks which sound almost as good as biamped, the latter is still preferable.

Figure 1.128 shows the groundplane measurements of the 8" subwoofer and the two-way prototype speaker. You could probably eliminate the 1dB sag by adjusting the crossover level and frequency controls. Figure 1.129 shows the full-range response with an extended and reasonably flat bandwidth.

A Final Note on Tweeter Variations

After I had finished the final prototype and prepared its measurements I received two new D2905 samples from Scan-Speak's managing director, Torben Sondergarrd. The new version (developed in conjunction with Proac) with differences primarily in the damping material between the dome and pole piece, required a revision in the high-pass crossover.

The D2905 used for the prototype was made in 1991. It has a hard damping material barely 1/8" beneath the dome surface, which causes the high end to be somewhat down at 20kHz. The new version uses a fluffy Dacron™-like material, and you can see the pole piece through the cloth.

To identify the version you have, either visually inspect the damping material beneath the dome (carefully push in on the dome to confirm the type of material used) or read the dome's label, which has a three-digit number indicating the week and year of manufacture. For example, 201 indicates the tweeter was manufactured during the 20th week of 1991; 1992 versions have numbers ending in 2.

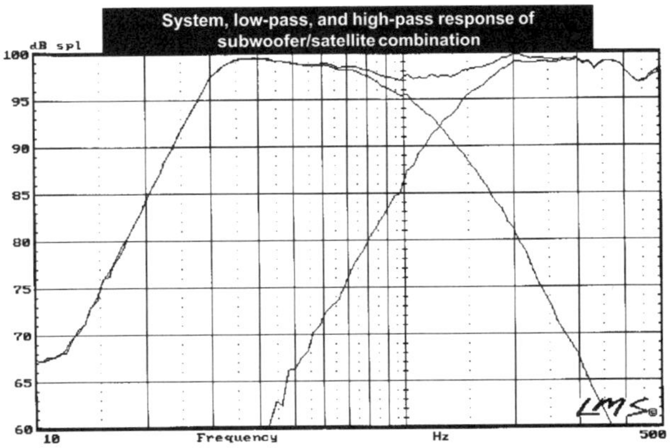

M1
M1 sub
M1 sat

FIGURE 1.128

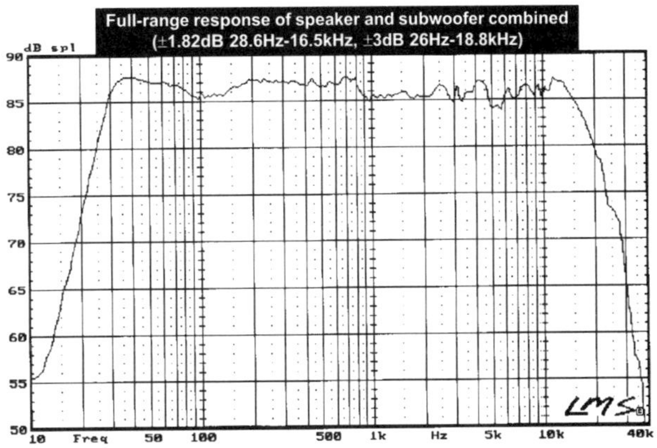

M1 + sub

FIGURE 1.129

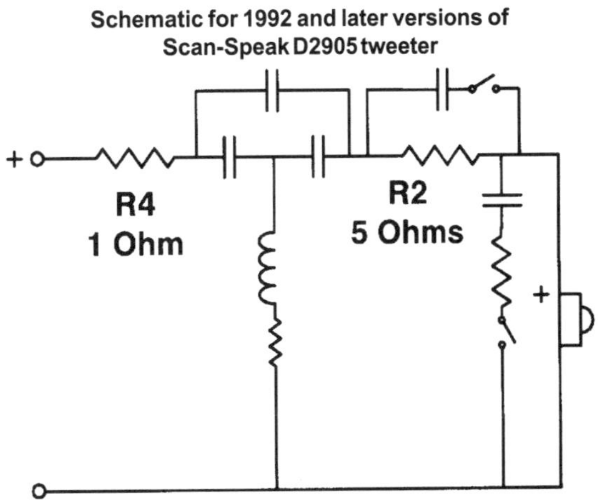

FIGURE 1.130

Although the newer tweeter measures better above 15kHz, both versions perform similarly. Figure 1.130 gives the high-pass crossover changes.

References

1. Dickason, V., The Loudspeaker Design Cookbook, Fourth Edition, 1991, Audio Amateur Press, p. 125 ff.

2. Colloms, M., High Performance Loudspeakers, 1978, 1985, 1991, Pentech Press, p. 225.

Chapter Two

A SEALED TWO-WAY

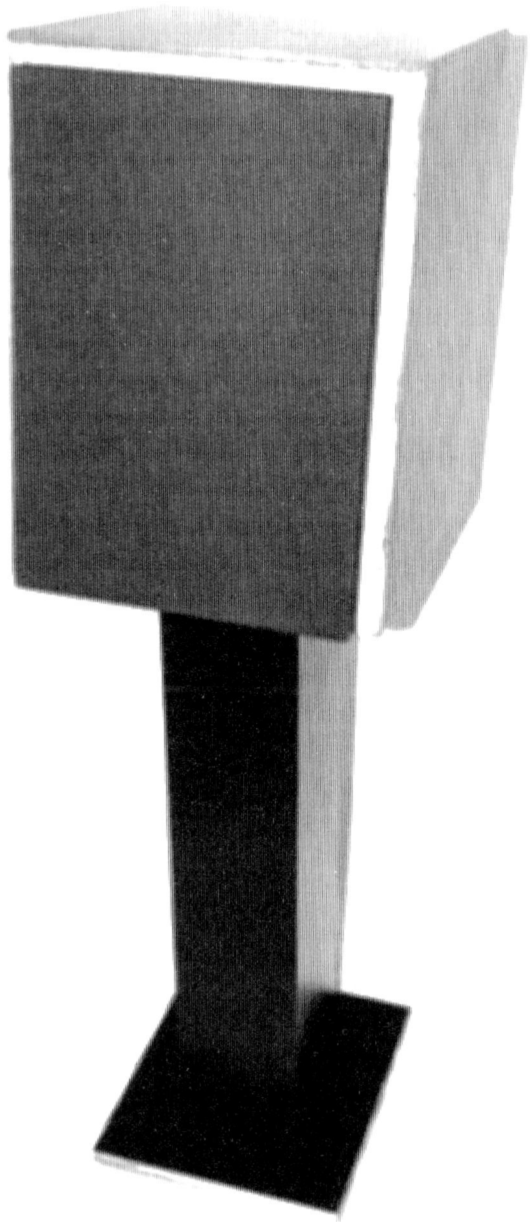

2.10 Overview

The second two-way design utilizes an 8" woofer and a 1" dome tweeter. As mentioned in Chapter 1, this combination is frequently regarded as the ultimate configuration for a two-way design: the former is capable of good low-end performance with adequate dynamic range, while the latter, if well-designed, can generally be coaxed into functioning at a sufficiently low crossover frequency. Used together, the result is a reasonably flat off-axis response window. For this project, I selected the Audio Concepts, Inc.(ACI) AC8 woofer and the Vifa D25AG-35-06 metal-dome tweeter.

In Chapter 1, I discussed many of the procedural details which accompanied the development of the four designs described in this book, and generally will not repeat them for the remaining designs. Chapters 2-4, for example, will not repeat the procedure used for measuring driver parameters, only list the actual calculated parameters that resulted from the procedure.

2.20 The Woofer

There always seems to be a scarcity in the OEM marketplace of good, well-damped woofers designed for sealed-box performance. For the past several years, however, Audio Concepts has been offering a number of well-designed drivers (8", 10", and 12") with parameters which are ideally suited for sealed enclosures. The AC8 is constructed with a stamped basket, poly cone, rubber surround, and an elevated spider for maximum linearity. An aluminum former is used for the two-layer, 40mm voice coil. ACI rates the speaker at 90dB SPL sensitivity, with an RMS power-handling rating of 100W.

After an appropriate break-in period, I tested the samples, and the T/S parameters generated as follows (the free-air impedance from which most of the data is derived is shown in Fig. 2.1):

	A	B
Z_{NOM}	8	8
R_E (Ω)	5.12	5.33
S_d (M^2)	0.0206	0.0206
Bl (TM)	7.75	7.92
V_{AS} (ltr.)	57.1	53.1
C_{MS} (µM/N)	947	881
f_S (Hz)	34.4	34.7
Q_{MS}	2.506	2.439
Q_{ES}	0.413	0.426
Q_{TS}	0.355	0.362
SPL (2.83V)	90.8dB	90.7dB
L_E (1kHz)	0.84mH	0.88mH
(20kHz)	0.29mH	0.31mH

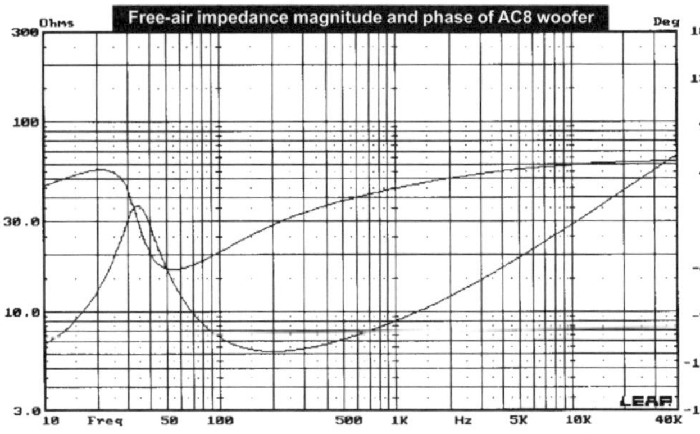

— AC8 free air

FIGURE 2.1

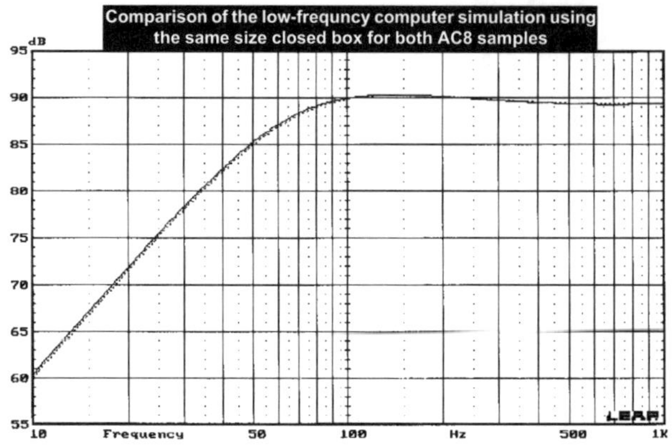

— AC8 sample A
······· AC8 sample B

FIGURE 2.2

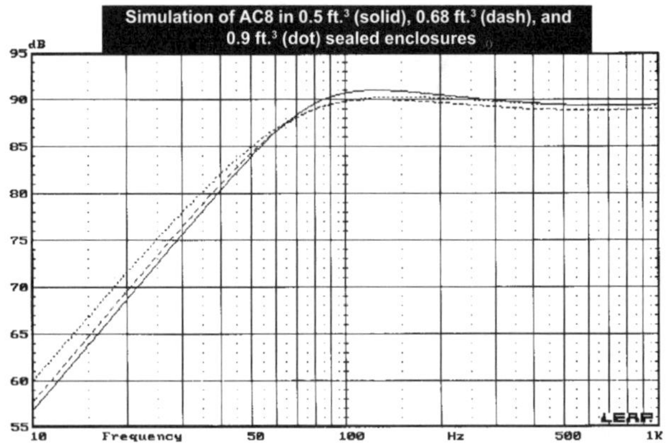

— AC8 .5cf f3=68Hz
······· M4 .9cf 81deg f3=63Hz 1W
- - - M4 .68cf 89deg f3=63Hz 1W

FIGURE 2.3

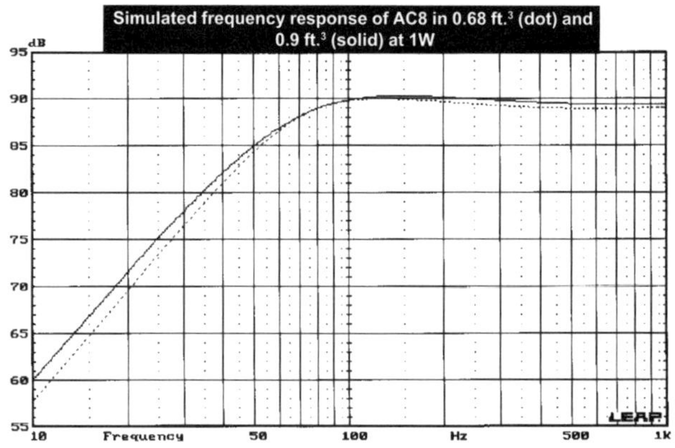

— M4 .9cf 81deg f3=63Hz 1W
······· M4 .68cf 89deg f3=63Hz 1W

FIGURE 2.4

The T/S parameters for the two samples are quite close, as is evident in the computer simulation in Fig. 2.2, which compares both in the same volume enclosure. Using the Quick Cabinet utility in LEAP, I determined that the AC8 would produce a Q_{TC}, or system Q, of 0.7 in a 0.68 ft.^{3} enclosure. This agrees with ACI's recommendation of from 0.7 to 1.2 ft.^{3} of enclosure volume for this woofer.

To further investigate possible enclosure configurations, I generated the response curves in Fig. 2.3 to compare three sealed-box simulations for the AC8. The box sizes I used were 0.5, 0.68, and 0.9 ft.^{3}, which varied somewhat from the initial approximation. For these simulations, the f_3s vary between 68Hz for the smallest-sized box to 63Hz for the two larger-sized boxes. To determine the box Qs, I measured the phase angle of the -3dB point on each curve using the graph cursor in LEAP. The resulting phase angles were 91° for the 0.5 ft.^{3} box (Q_{TC} = 0.75), 0.7 ft.^{3} at 89.5° (Q_{TC} = 0.7), and 0.9 ft.^{3} at 81° (Q_{TC} somewhat less than 0.65). Apparently, changes in the total box volume have only relatively minor influence on this woofer's f_3 and Q_{TC} for box sizes larger than 0.75 ft.^{3}

After successfully using the semicylindrical internal enclosure in Chapter 1's speaker, I decided to use the same technique with both this speaker and the one in Chapter 3. Given the diameter of the driver, however, it was impractical to use the same 8-inch-diameter Sonotube as in the first speaker. I would need to use the next larger size, which is 10" in diameter. Since the Sonotube has a " wall, this dictated an internal box width of 10.5". This dimension, plus the vertical height needed to mount both drivers, determined the final 0.9 ft.^{3} box volume.

Before building the prototype enclosure, I performed a LEAP simulation to compare the performance at 1W and at a predetermined higher power input level (sufficient to increase cone excursion to X_{MAX}) for both the original calculated size of 0.68 ft.^{3} and the proposed 0.9 ft.^{3} enclosure. Figures 2.4-2.7 show the comparison

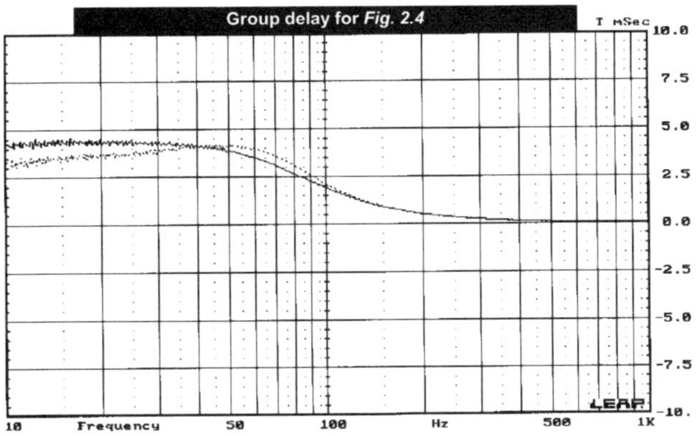

— M4 .9cf 81deg f3=63Hz 1W
······· M4 .68cf 89deg f3=63Hz 1W

FIGURE 2.5

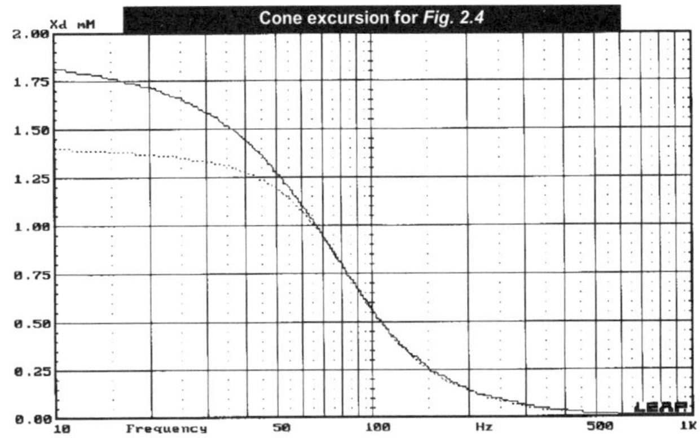

— M4 .9cf 81deg f3=63Hz 1W
······· M4 .68cf 89deg f3=63Hz 1W

FIGURE 2.6

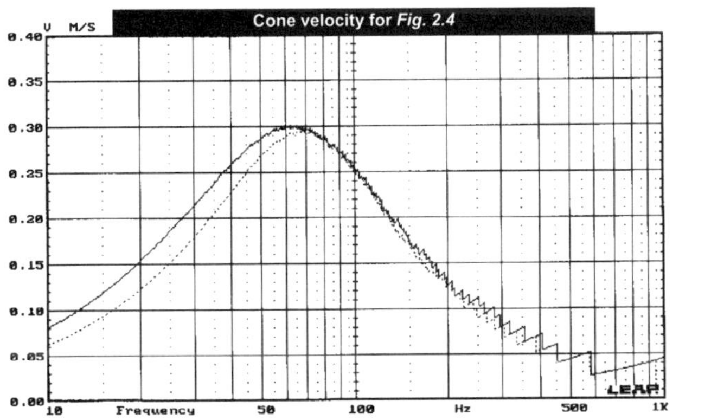

— M4 .9cf 81deg f3=63Hz 1W
······· M4 .68cf 89deg f3=63Hz 1W

FIGURE 2.7

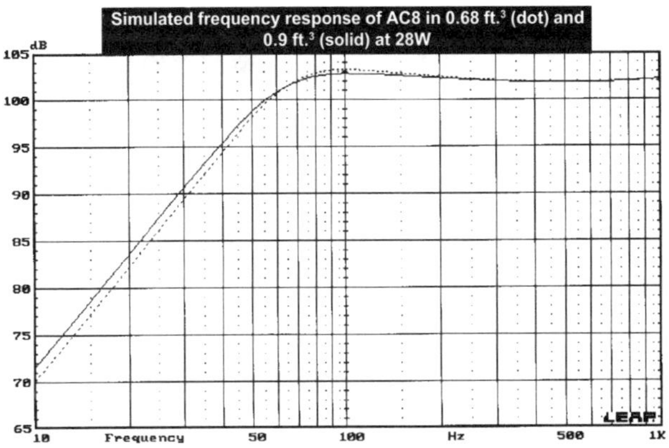

— M4 .9cf 95deg f3=55Hz 28W
······· M4 .68cf 98deg f3=59Hz 28W

FIGURE 2.8

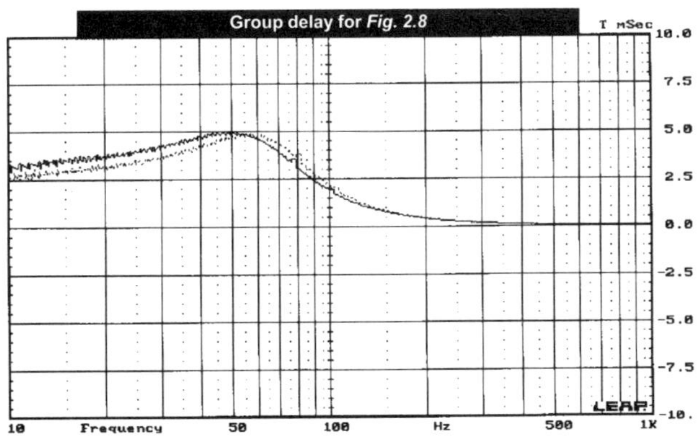

— M4 .9cf 95deg f3=55Hz 28W
······· M4 .68cf 98deg f3=59Hz 28W

FIGURE 2.9

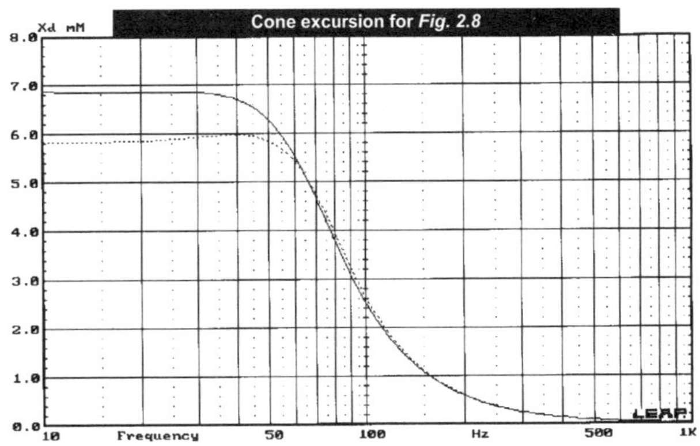

— M4 .9cf 95deg f3=55Hz 28W
······· M4 .68cf 98deg f3=59Hz 28W

FIGURE 2.10

of both enclosure volumes at the 1W input level for frequency response, group delay, cone excursion, and cone velocity, respectively. The group delay curves are typical of a critically damped, sealed enclosure.

The same curves are shown in Figs. 2.8-2.11 for a 28W input power level. The cone excursion curves in Fig. 2.10 show that the larger 0.9 ft.^{3} enclosure's maximum excursion is still slightly less than 7mm with an indicated SPL of 103dB (Fig. 2.8). Since the AC8's X_{MAX} is 6.5mm, the speaker is well within the X_{MAX} + 15% criteria, which would specify a 7.5mm threshold for a 3% harmonic distortion level. The AC8 thus will

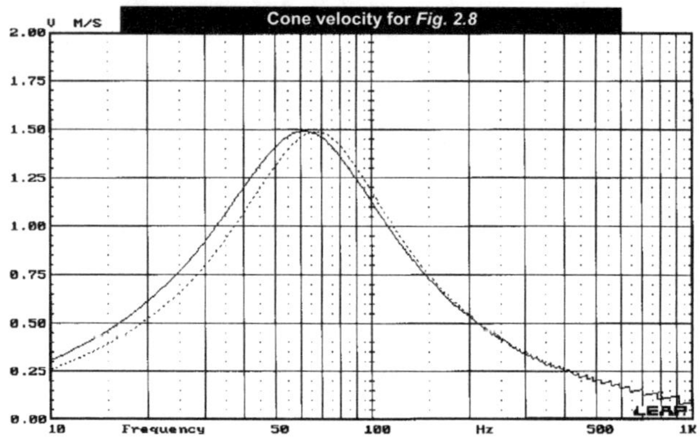

— M4 .9cf95deg f3=55Hz 28W
······· M4 .68cf98deg f3=59Hz 28W

FIGURE 2.11

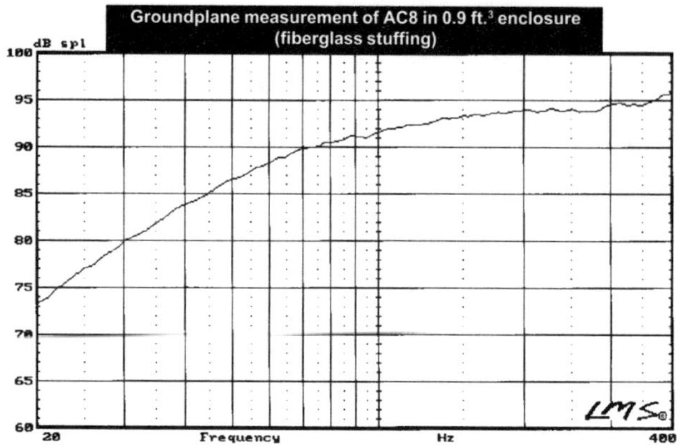

— Wofb qp

FIGURE 2.12

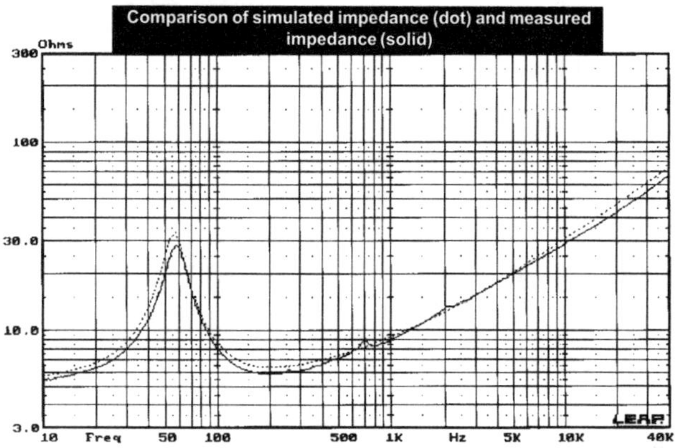

— AC8 in box foam in as
······· M4 .9cf 81deg f3=63Hz 1W

FIGURE 2.13

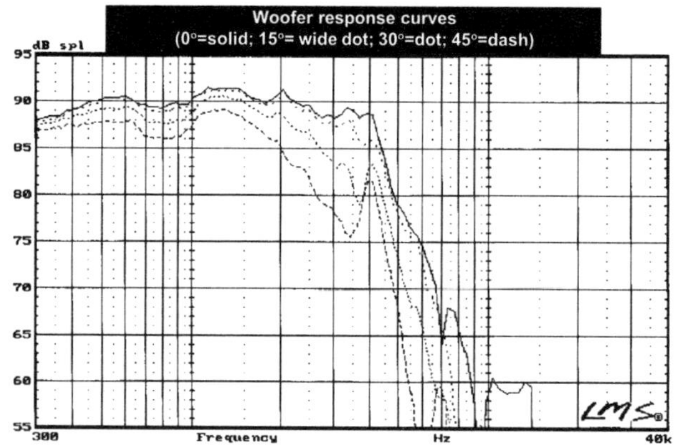

- - - Wof a15
— Wof a0 6
······· Wof a30
--- Wof a45

FIGURE 2.14

provide linear low-distortion performance up to a very loud 103dB+ SPL—very adequate for a single home speaker.

Once the prototype was built, I damped the enclosure by lining the rear of the cardboard semicylinder and all four walls with acoustic foam (the egg-crate type sold by ACI as "AC Foam"). I damped the area directly behind the woofer (the bottom half of the enclosure) with slightly teased Acousta-Stuf damping material. Once these materials were in place, I measured and calculated the woofer's final Q_{TC} to be exactly as the simulation predicted: 0.655.

A groundplane measurement verified the response, as depicted in Fig. 2.12. (This preliminary measurement was made with fiberglass stuffing material, which yielded a much lower Q.) Since this measurement includes the baffle diffraction, it is difficult to calculate the speaker's f_3; however, knowing that the rise begins close to 100Hz for this size baffle, the f_3 would be close to 65Hz. The shallow rolloff produced by the low-Q sealed box may not seem sufficient to produce impressive bass, but it actually performs well in small rooms. Typical listening rooms tend to exhibit a lift or degree of room gain which adds about 3dB at 80Hz and 9dB at 20Hz, so the response of this type of woofer seems much less anemic than implied by the response curve. Figure 2.13 compares the simulated impedance with the measured impedance of the woofer mounted in the box.

The woofer's on- and off-axis response (0°, 15°, 30°, and 45°) is illustrated in Fig. 2.14. At 30°, the response is about 2.5dB down from the 0° curve at 2kHz, and is 5-6dB down at 3kHz. Although an ideal crossover point for this woofer would be 1.5kHz, 2kHz represents a good compromise. This accommodates the excursion capability of a 1" dome tweeter with a low-resonance point, such as the Vifa D25AG, and still elicits good off-axis response from the woofer.

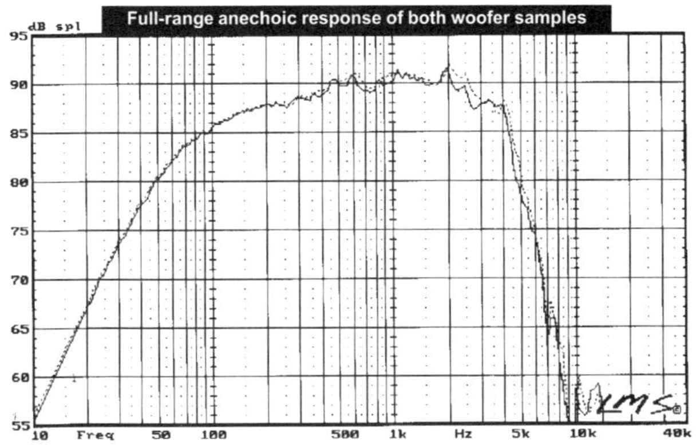

— Wof a fr
········ Wof b fr

FIGURE 2.15

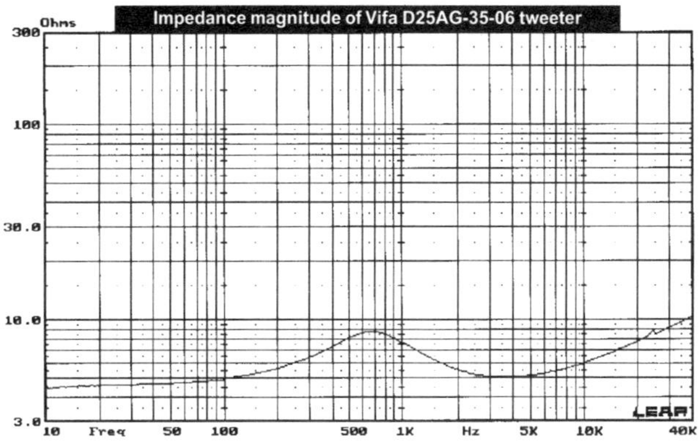

— Vifa D25AG no xover

FIGURE 2.16

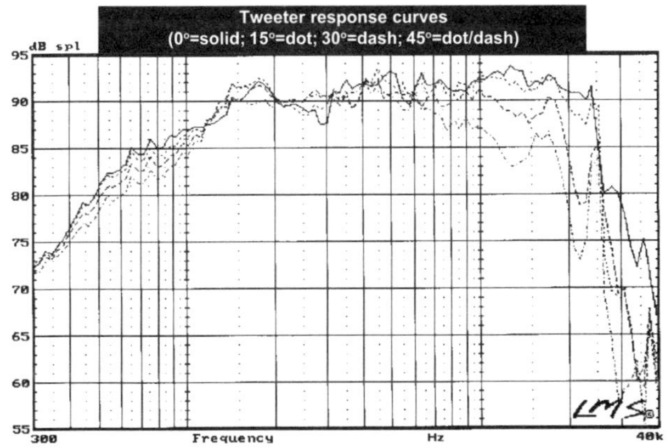

— Tw a0 6
········ Tw a15
- - - Tw a30
-·-· Tw a45

FIGURE 2.17

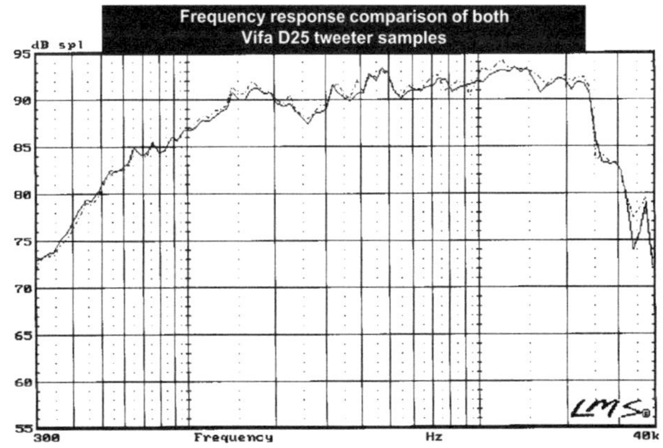

— Tw 0b vert
-·-· Tw a0 vert

FIGURE 2.18

Figure 2.15 gives the full-range anechoic response of both woofer samples (made by splicing the groundplane measurement with the gated-tower measurement). These drivers are fairly well-matched, exhibiting only a 1dB difference between 2-3kHz. Given the typical gain of most rooms, this speaker can be easily targeted for a nominal 87-88dB efficiency level.

2.30 The Tweeter

Vifa's D25AB-35-06, released in April 1992, is an excellent tweeter, and is even more spectacular considering its moderate price. The device is comprised of a 1" aluminum-alloy dome, with a vented pole piece and damped cavity. Figure 2.16 gives its impedance; although damped with magnetic fluid, the resonance rises 3.8Ω above the 5Ω impedance minimum. I exported the LMS-generated impedance plot into LEAP and measured the parameters, which yielded the following:

Z_{NOM}	6Ω
R_E	4.6Ω
f_S	668Hz
Q_{MS}	0.765
Q_{ES}	0.857
Q_{TS}	0.404
L_E (1kHz)	0.088mH
(20kHz)	0.031mH

The responses at 0°, 15°, 30°, and 45° are shown in Fig. 2.17. Above 2kHz, the on-axis response is ±2.5dB out to 25kHz, with a smooth off-axis response out to 45°. Both samples were well-matched to ±0.5dB, as depicted in Fig. 2.18.

2.40 The Crossover Design

The physical time delay between the Vifa D25AG and the AC8, mounted flush on the same baffle and measured frontplate-to-frontplate, is 2.17". This represents 0.375" (the distance from the baffle to

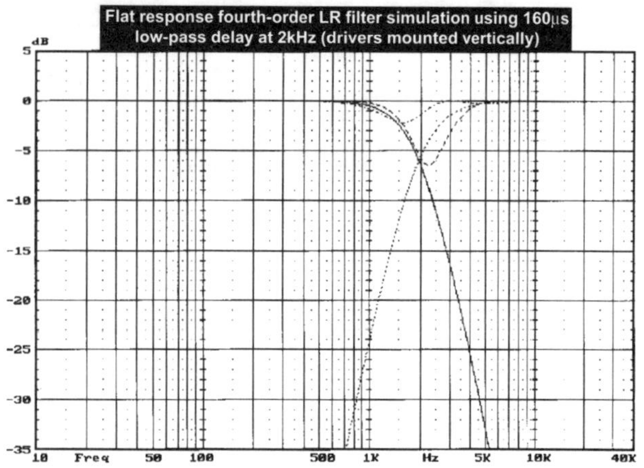

—— Lo Pass Section Only: Magnitude
········ Hi Pass Section Only: Magnitude
– – – Sum=Lo+Hi (InPhase): Magnitude
–·–· Sum=Lo-Hi (Reverse): Magnitude

Lo Filtr: 4th-LnkRly	Hi Filtr: 4th-LnkRly
Freq = 2000.00 Hz	Freq= 2000.00 Hz
Level= 0.00 dB	Level= 0.00 dB
Delay= 160.00µS	Delay= 0.00 µS

FIGURE 2.19

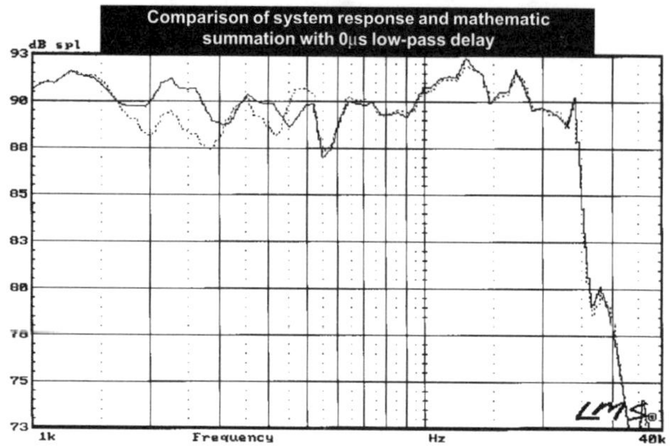

—— Sys 0 sw
········ Sum 0 delay

FIGURE 2.20

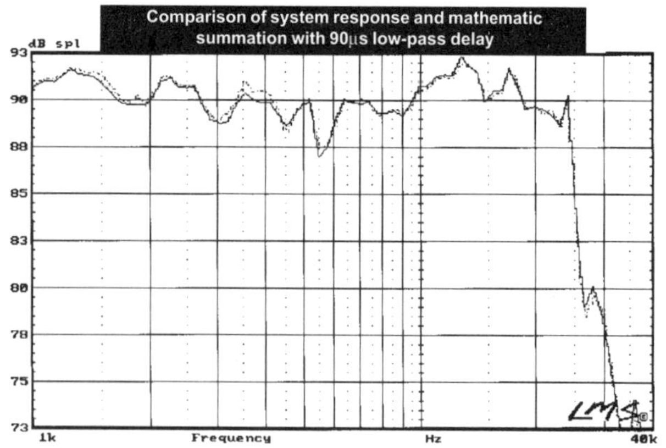

—— Sys 0 sw
–·–· Sum 90uS

FIGURE 2.21

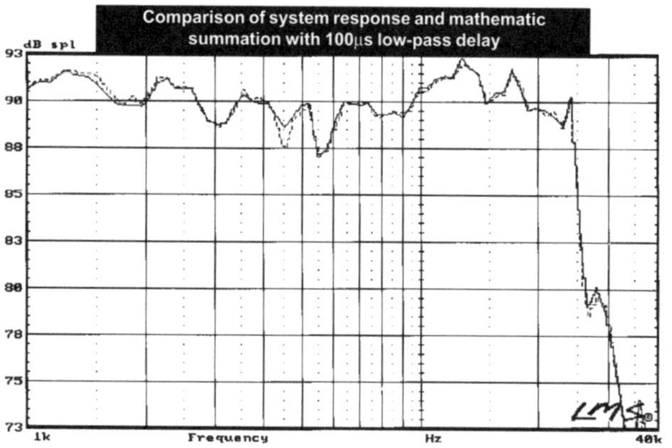

—— Sys 0 sw
– – – Sum 100uS

FIGURE 2.22

the center of the tweeter frontplate) subtracted from 2.54" (the distance from the baffle to the center of the woofer frontplate). Expressed in terms of time interval, 2.17" equals 160µs.

Figure 2.19 shows the simulated summation of two acoustically flat, fourth-order L-R high- and low-pass sections, with the latter displaced by 160µs at a crossover frequency of 2kHz. The in-phase summation is 6dB down at 2.2kHz, and the out-of-phase summation is 2dB down at 1.5kHz, indicating that the drivers are close to being acoustically out-of-phase with an in-phase electrical connection. Since 160µs is quite a lot of delay for which to compensate, I decided to also tilt this speaker's baffle at the same modest 6° slope. This should cut about 60µs from the delay time, but not radically alter the drivers' on-axis response.

I confirmed this using the method described in Chapter 1. The graphs in Figs. 2.20-2.23 show the iterated delay process for 0, 90, 100, and 110µs of delay, respectively. This display depicts a greater bandwidth than that shown in Chapter 1, for a better idea of how the phase picture changes and tilts from low to high frequency as the delay is increased. The plot in Fig. 2.22 establishes that, with the baffle sloped at 6°, the actual amount of time delay between the two drivers is 100µs.

With the physical time delay reduced to 100µs, the relationship of summing two fourth-order sections at 2kHz changes considerably, as seen in

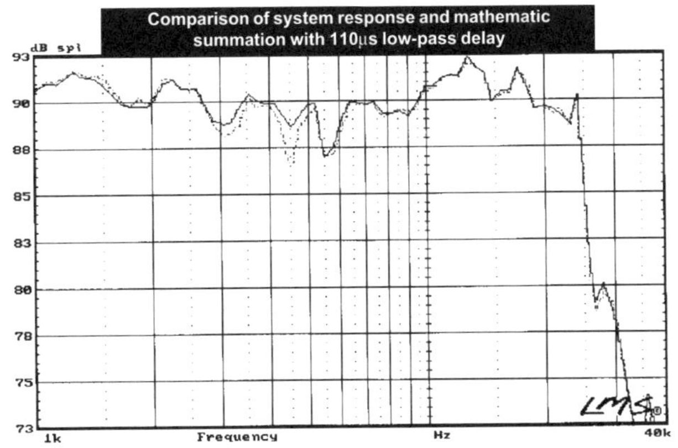

— Sys 0 sw
-·-· Sum 110uS

FIGURE 2.23

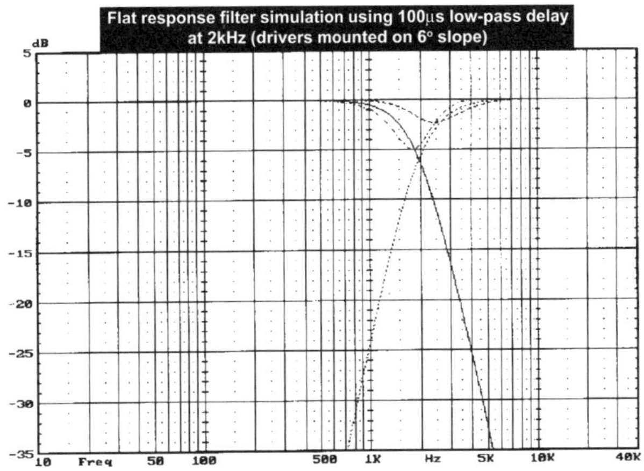

— Lo Pass Section Only: Magnitude
······· Hi Pass Section Only: Magnitude
- - - Sum=Lo+Hi (InPhase): Magnitude
-·-· Sum=Lo-Hi (Reverse): Magnitude

Lo Filtr: 4th-LnkRly	Hi Filtr: 4th-LnkRly
Freq = 2000.00 Hz	Freq= 2000.00 Hz
Level= 0.00 dB	Level= 0.00 dB
Delay= 100.00µS	Delay= 0.00 µS

FIGURE 2.24

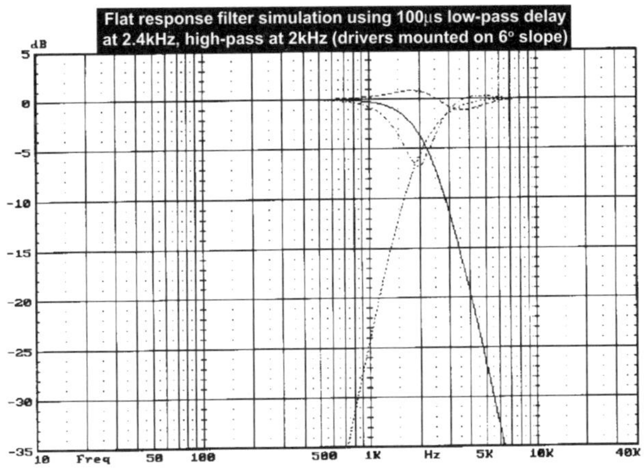

— Lo Pass Section Only: Magnitude
······· Hi Pass Section Only: Magnitude
- - - Sum=Lo+Hi (InPhase): Magnitude
-·-· Sum=Lo-Hi (Reverse): Magnitude

Lo Filtr: 4th-LnkRly	Hi Filtr: 4th-LnkRly
Freq = 2400.00 Hz	Freq= 2000.00 Hz
Level= 0.00 dB	Level= 0.00 dB
Delay= 100.00µS	Delay= 0.00 µS

FIGURE 2.25

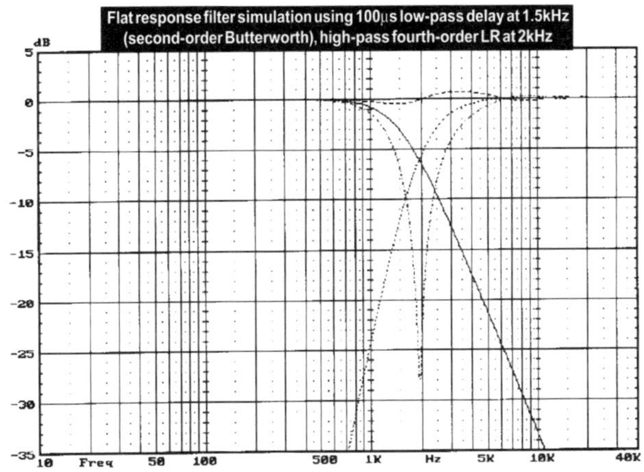

— Lo Pass Section Only: Magnitude
······· Hi Pass Section Only: Magnitude
- - - Sum=Lo+Hi (InPhase): Magnitude
-·-· Sum=Lo-Hi (Reverse): Magnitude

Lo Filtr: 4th-LnkRly	Hi Filtr: 4th-LnkRly
Freq = 1500.00 Hz	Freq= 2000.00 Hz
Level= 0.00 dB	Level= 0.00 dB
Delay= 100.00µS	Delay= 0.00 µS

FIGURE 2.26

Fig. 2.24. The in-phase summation is now down only 2dB. The SPL variation in the crossover frequency region ideally should be no greater than 1dB, however, so I determined two methods for creating a flatter-magnitude response. The first, illustrated in *Fig. 2.25*, was derived by overlapping the crossover frequencies of a fourth-order crossover. I set the high-pass section to 2kHz and increased the low-pass section frequency to 2.4kHz. The result is an acceptable 1dB ripple.

The alternative is to use an asymmetrical crossover and skew the network frequencies. *Figure 2.26* shows the results of summing a second-order Butterworth at 1.5kHz with a fourth-order L-R at 2kHz. Once again, the result is 0.5-1dB ripple. This analysis presents two viable crossover formats: the skewed symmetrical fourth-order and the skewed asymmetrical second-/fourth-order.

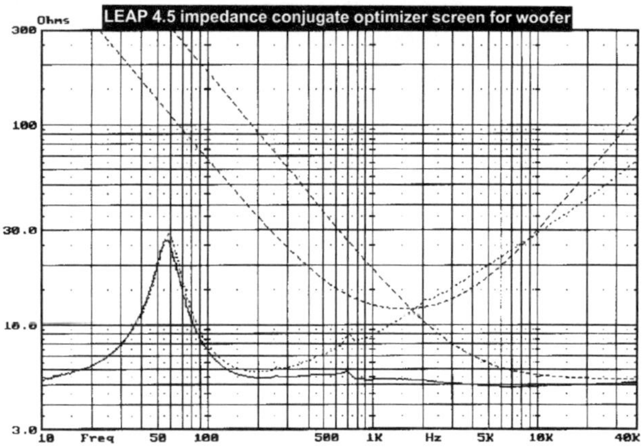

— Combined Impedance
······· Original Impedance
- - - Conjugate Nets

DGL Data Info
DGL# 1:AC8
Wof no xover

FIGURE 2.27

Woofer #1 low-pass filter schematic

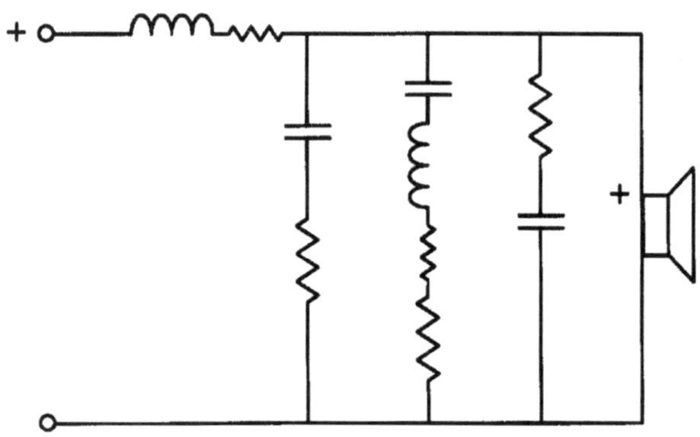

FIGURE 2.28

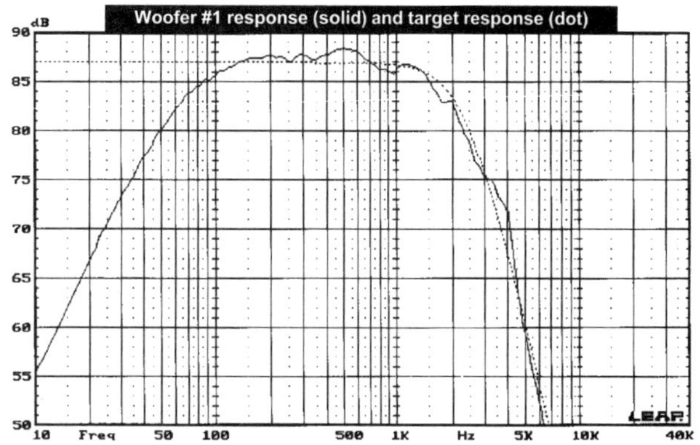

— Wof 2 2nd + conj 4th at 2400 p10
······· Generic 1p

FIGURE 2.29

2.50 The Woofer Crossover

From the woofer's full-range anechoic response in *Fig. 2.15*, we see that the AC8 exhibits the same rising response caused by cabinet diffraction similar to that seen in the Vifa P17WJ. The AC8 frequency response is rising about 4dB, from 200Hz to 2kHz. To compensate for this, the low-pass network must begin contouring the woofer's response at 200Hz by gradually decreasing the voltage drive from 200Hz to 2kHz, and then attenuating the response at a fourth-order rate.

The first step was to examine the possibility of using LCR and CR conjugates to alter the woofer's impedance to a flat nominal value. *Figure 2.27* depicts the LEAP Conjugate Designer screen, which yielded the following values:

LCR	CR
0.44mH	8.6μF
23.68μF	5.3Ω
12Ω	

With these values as a reference, I created a number of possible network scenarios. Due to the power of the LEAP software, within a three-hour time period I was able to design 18 different permutations of high- and low-pass sections using different component values and circuit topographies, and all producing a system response of at least ±2dB. Try doing that by trial and error! For the purposes of this book, I examined only four of the low-pass circuits: #1 is a fourth-order L-R set at 2.4kHz; #2 is a second-order Butterworth at 1.6kHz; #3 and #4 are second-order Butterworth sections at 1.5kHz. Their topographies are described as follows:

1. Second-order low-pass + LCR and CR conjugates
2. First-order low-pass + half-order + LCR and CR conjugates
3. Second-order low-pass + half-order + CR conjugate
4. First-order low-pass + CR conjugate

I derived their respective component values using the same manual iteration and computer optimization as used elsewhere throughout this book. The results of this process and schematic diagrams are illustrated by the following:

Woofer#	Circuit	Response +Target	Z	Transfer Function
1	2.28	2.29	2.30	2.31
2	2.32	2.33	2.34	2.35
3	2.36	2.37	2.38	2.39
4	2.40	2.41	2.42	2.43

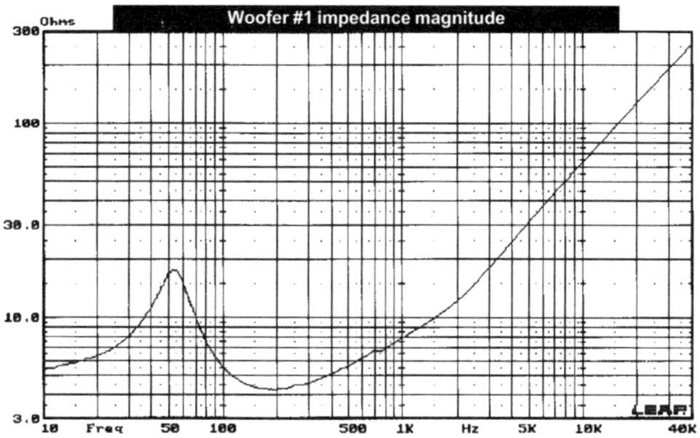

— Wof 2 2nd + conj 4th at 2400 p10

FIGURE 2.30

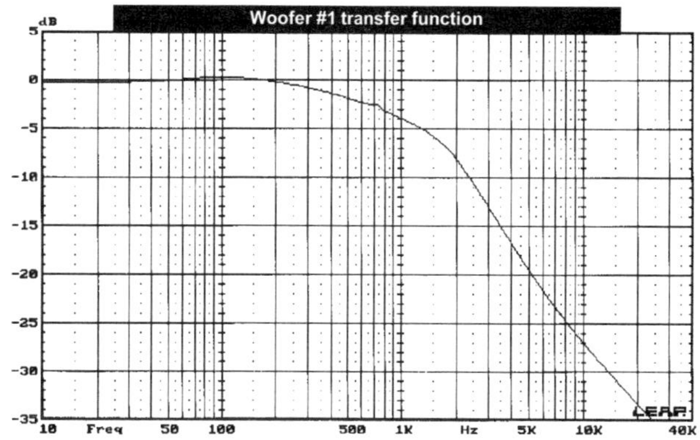

— Wof 2 2nd + conj 4th at 2400 p10

FIGURE 2.31

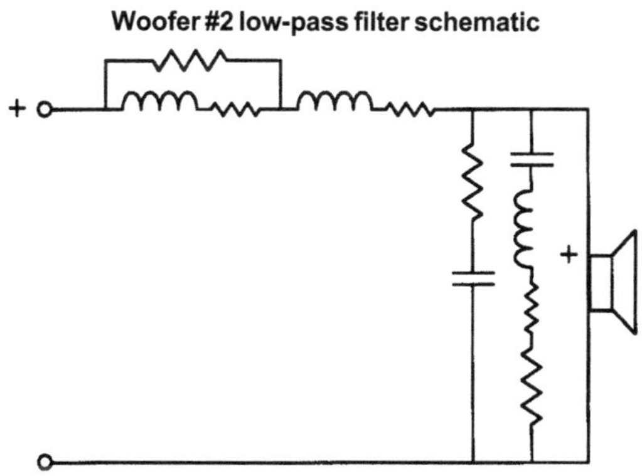

FIGURE 2.32

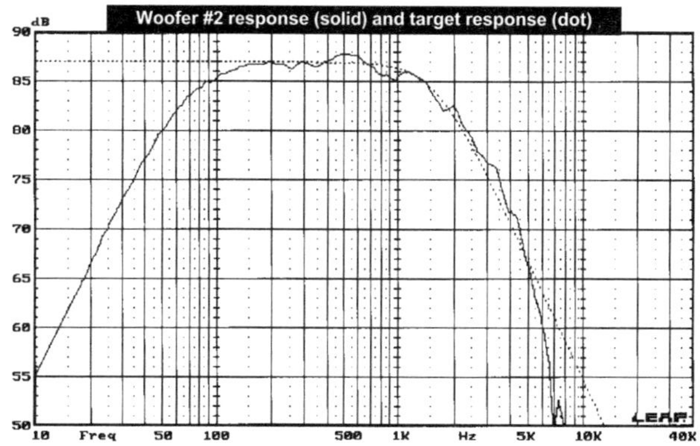

— Wof 6 2nd but at 1600 p7
······· Generic lp

FIGURE 2.33

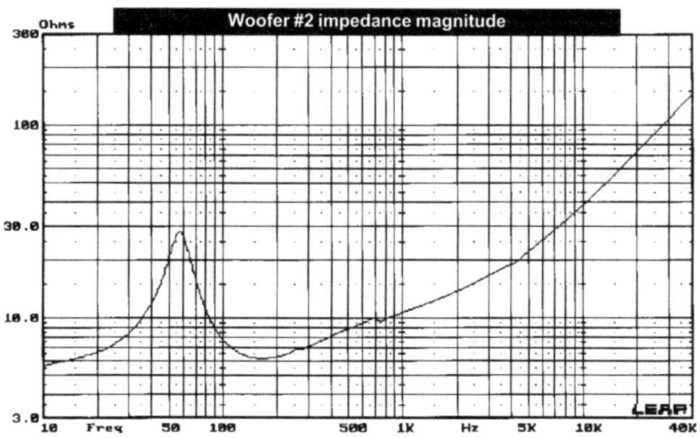

— Wof 6 2nd but at 1600 p7

FIGURE 2.34

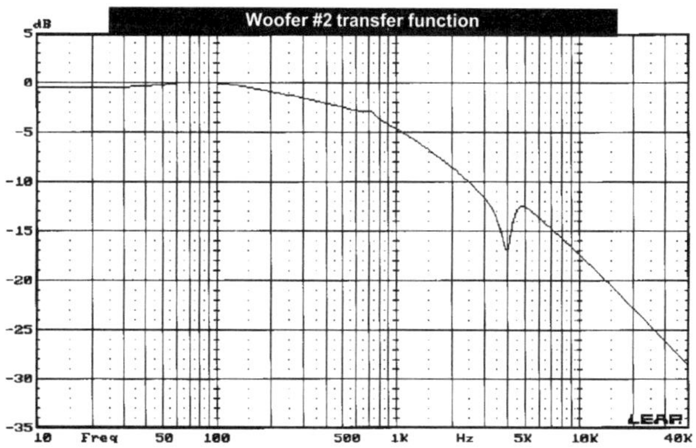

— Wof 6 2nd but at 1600 p7

FIGURE 2.35

All four circuits yielded usable transfer functions and impedance profiles. Their frequency responses, impedance and transfer functions are displayed simultaneously for comparison in *Figs. 2.44-2.46*. Even though each circuit optimized well to the targets, the required series inductances ranged from 1mH to as much as 2mH. Although the asymmetric crossover looked quite workable, I opted to use the skewed fourth-order circuit #1, primarily because it required only 1mH of series inductance, and I wanted to keep insertion losses from air-core inductors as low as possible. As with the circuit used with the P17WJ, the LCR causes the impedance minimum to be low enough to preclude paralleling two or more speakers together on one amplifier channel.

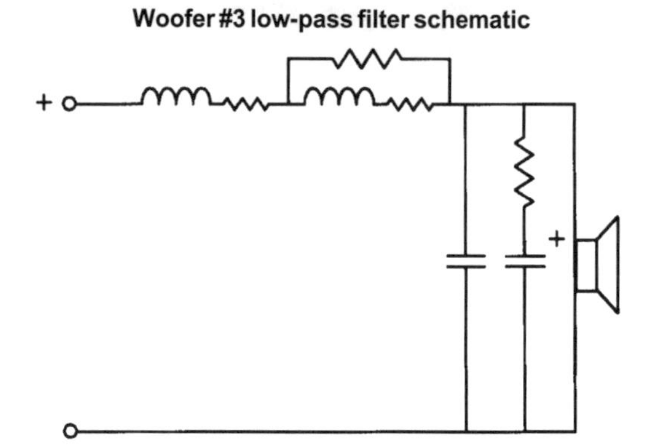

FIGURE 2.36

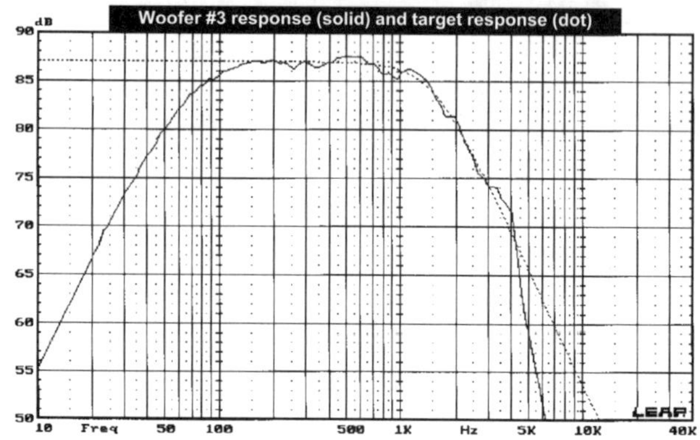

—— Wof 3 2nd but at 1500 p4
········ Generic 1p

FIGURE 2.37

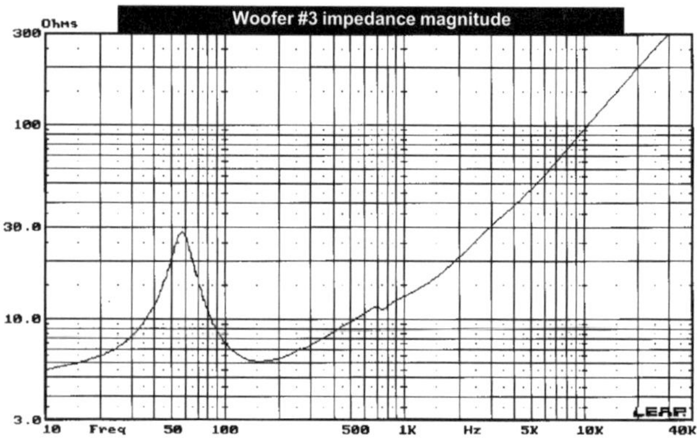

—— Wof 3 2nd but at 1500 p4

FIGURE 2.38

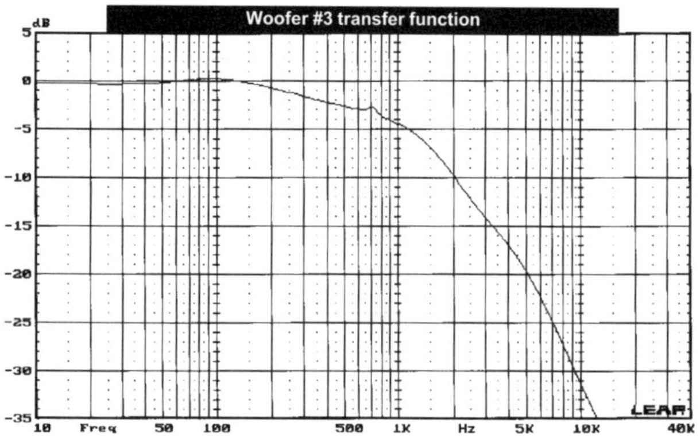

—— Wof 3 2nd but at 1500 p4

FIGURE 2.39

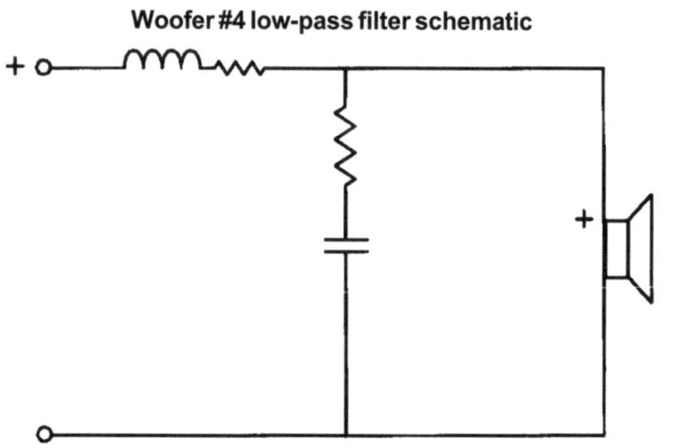

FIGURE 2.40

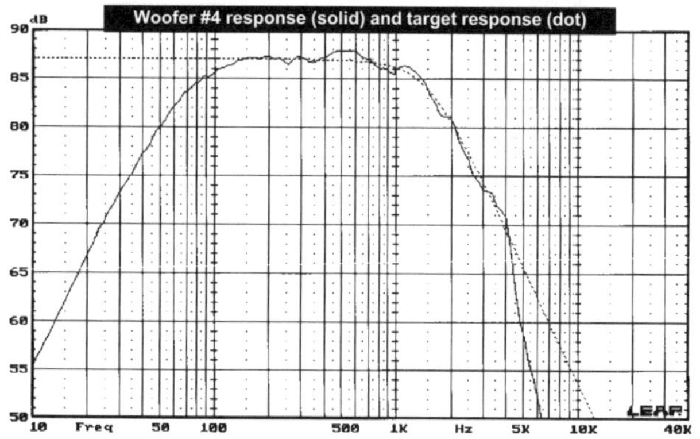

—— Wof 1 2nd but at 1500 p2
········ Generic 1p

FIGURE 2.41

2.60 The Tweeter Crossover

Due to the magnetic fluid in the D25AG's gap, and the resultant small differential between the impedance at the crossover region and the resonance, I didn't incorporate an LCR conjugate into any of the tweeter circuits (although that is a possibility). The tweeter's response in *Fig. 2.17* shows an approximate second-order rolloff below 1.8kHz and an SPL that averages around 91-92dB. Since the woofer's SPL target has been determined to be 87dB, the tweeter must be attenuated by about 4-5dB.

Using the 87dB target SPL, I optimized three different tweeter high-pass topographies to a fourth-order L-R target response at 2kHz. The circuit descriptions are:

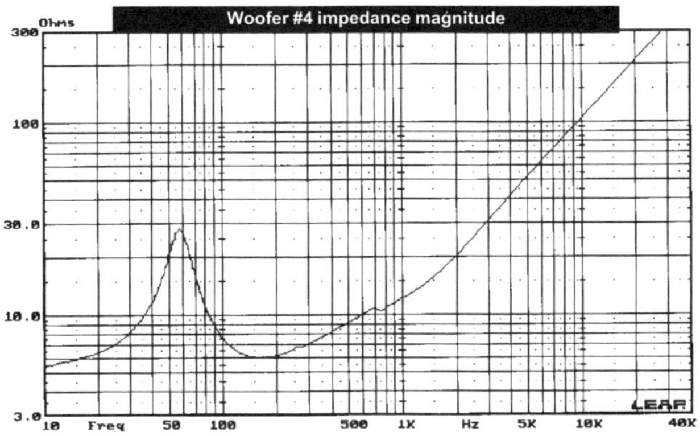

— Wof 1 2nd but at 1500 p2

FIGURE 2.42

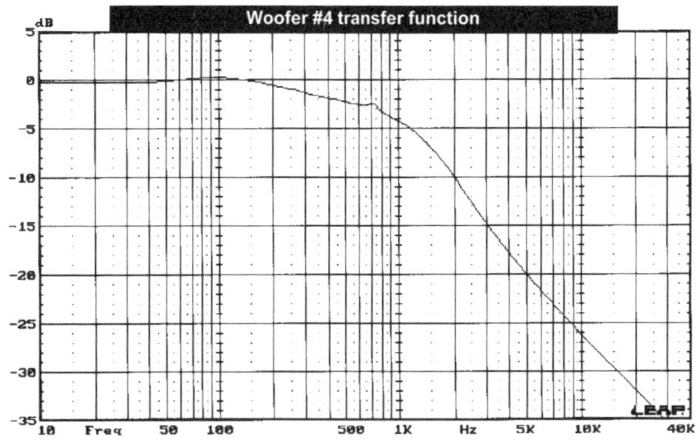

— Wof 1 2nd but at 1500 p2

FIGURE 2.43

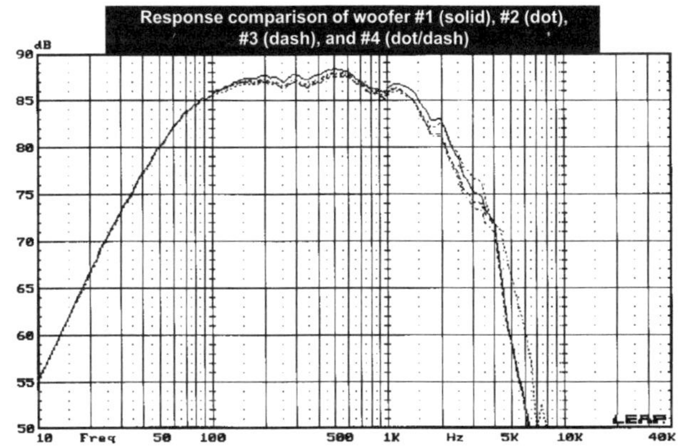

— Wof 2 2nd + conj 4th at 2400 p10
······ Wof 6 2nd but at 1600 p7
--- Wof 3 2nd but at 1500 p4
-·-· Wof 1 2nd but at 1500 p2

FIGURE 2.44

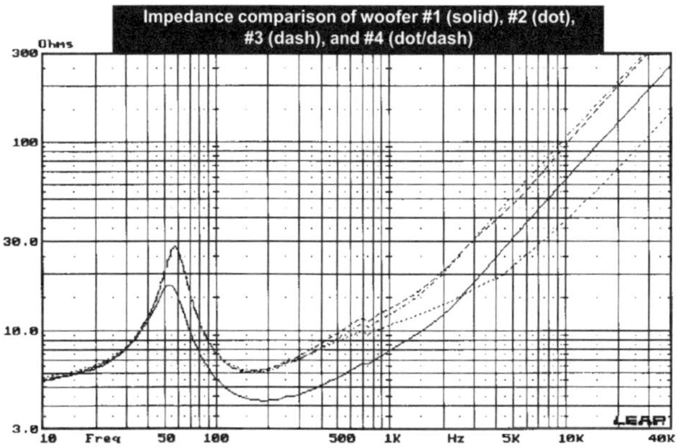

— Wof 2 2nd + conj 4th at 2400 p10
······ Wof 6 2nd but at 1600 p7
--- Wof 3 2nd but at 1500 p4
-·-· Wof 1 2nd but at 1500 p2

FIGURE 2.45

1. Second-order high-pass + series attenuation resistor + CR conjugate.
2. Third-order high-pass + series attenuation resistor.
3. Third-order high-pass + L-pad attenuation.

The following diagrams illustrate the optimized high-pass sections:

Tweeter#	Circuit	Response +Target	Z	Transfer Function
1	2.47	2.48	2.49	2.50
2	2.51	2.52	2.53	2.54
3	2.55	2.56	2.57	2.58

Figures 2.59-2.61 simultaneously compare their responses, impedances and transfer functions. While all three circuits optimized close to the L-R target function, tweeter #1 did not track the target well in the second octave below the crossover frequency, as exhibited by the 10dB scale response graph in *Fig. 2.62*.

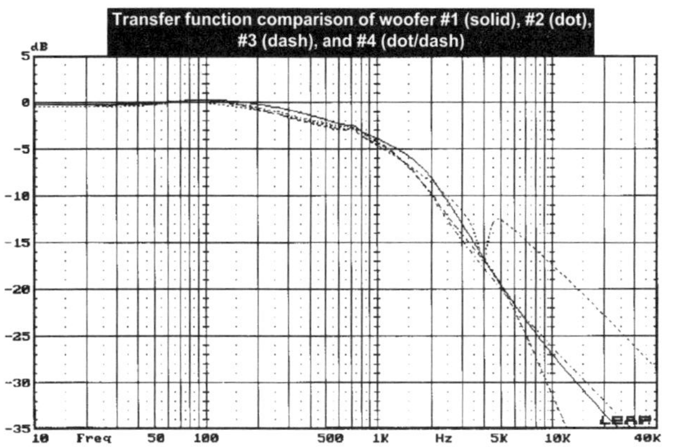

— Wof 2 2nd + conj 4th at 2400 p10
······ Wof 6 2nd but at 1600 p7
--- Wof 3 2nd but at 1500 p4
-·-· Wof 1 2nd but at 1500 p2

FIGURE 2.46

2.70 System Analysis

To choose which tweeter circuit to use with the woofer #1 network, I compared the summation of each pair of high- and low-pass sections. The three

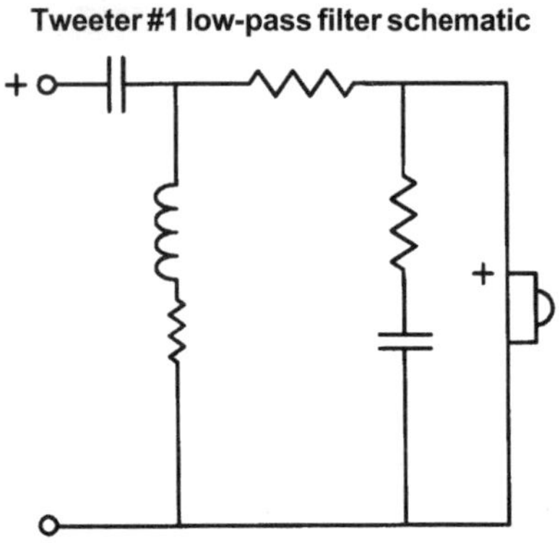

FIGURE 2.47

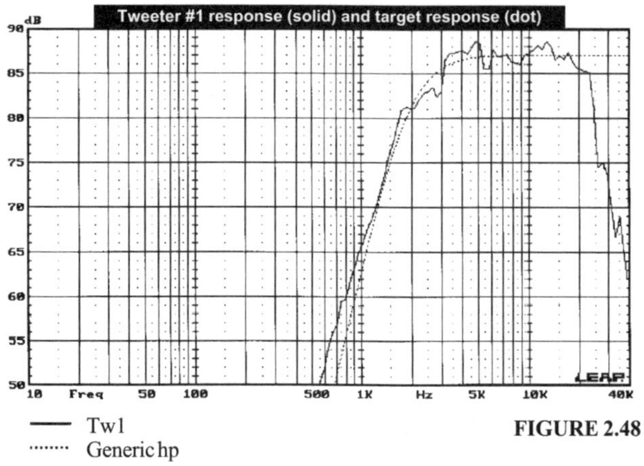

FIGURE 2.48

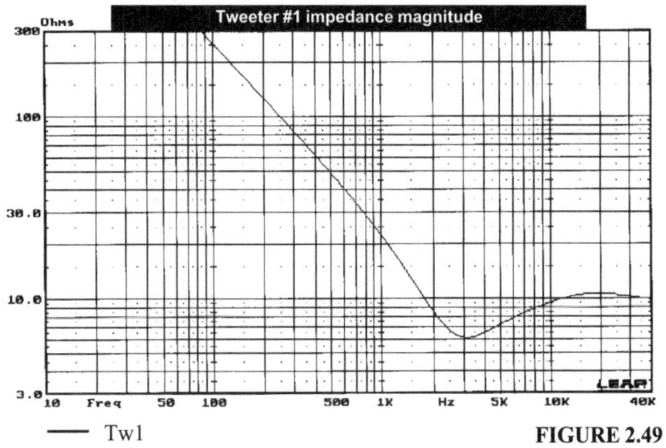

FIGURE 2.49

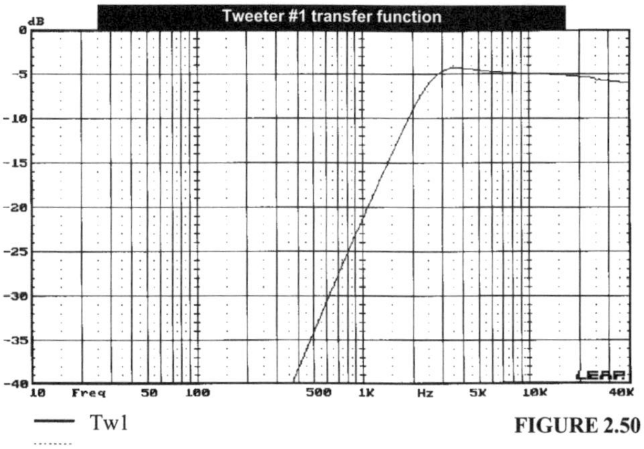

FIGURE 2.50

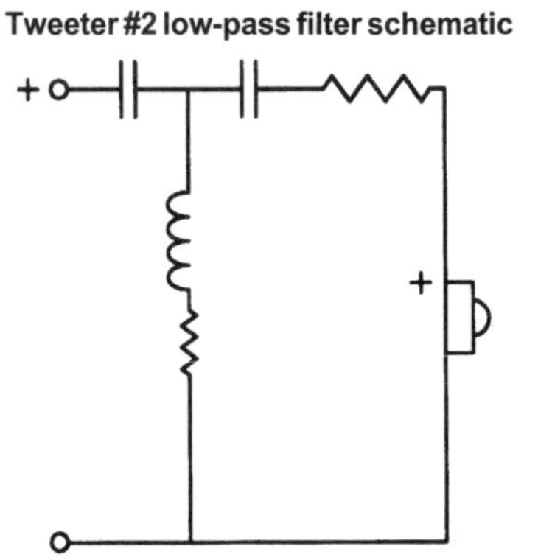

FIGURE 2.51

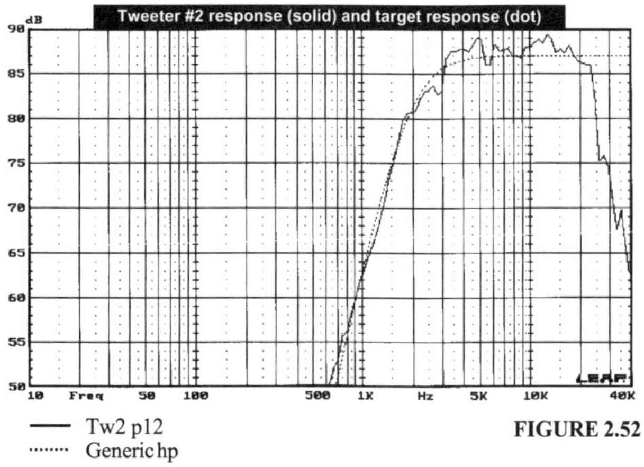

FIGURE 2.52

possible combinations are displayed with both the electrically in-phase summation and the polarity reversed (on the tweeter) out-of-phase summation:

Tweeter #	In-phase	Out-of-phase
1	2.63	2.64
2	2.65	2.66
3	2.67	2.68

The woofer #1/tweeter #1 combination's out-of-phase curve did not produce much of a null in the response, indicating the pair was less in-phase than the other combinations. Woofer #1/tweeter #2 and woofer #1/tweeter #3 produced reasonably deep nulls with the polarity reversed, although not as symmetrical as the speaker in *Chapter 1*. Comparing all three in-phase responses in *Fig.*

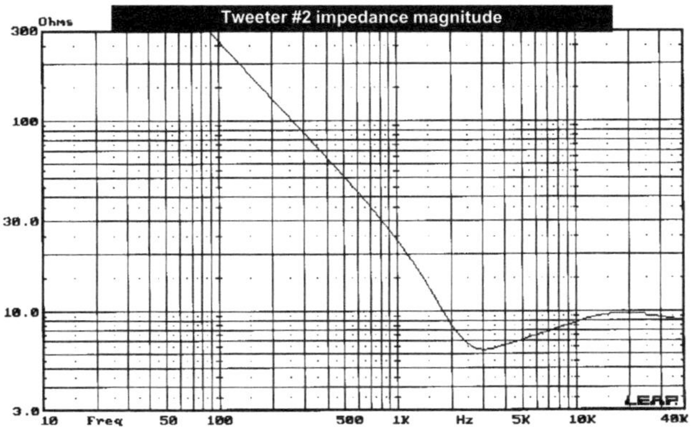

— Tw2 p12

FIGURE 2.53

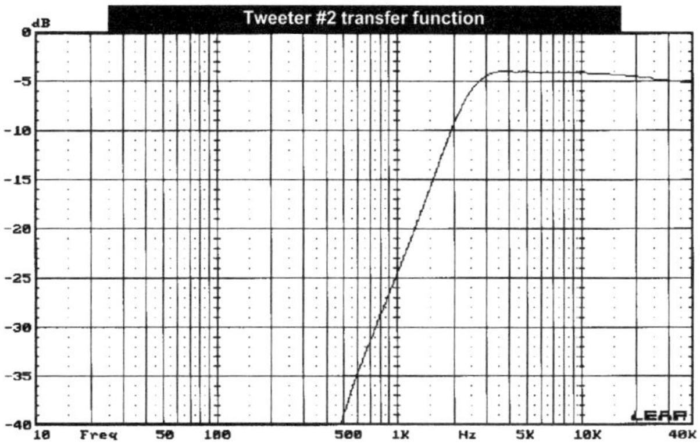

— Tw2 p12

FIGURE 2.54

Tweeter #3 low-pass filter schematic

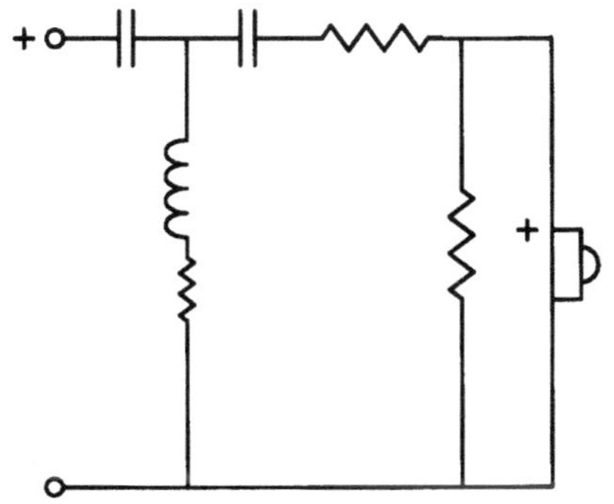

FIGURE 2.55

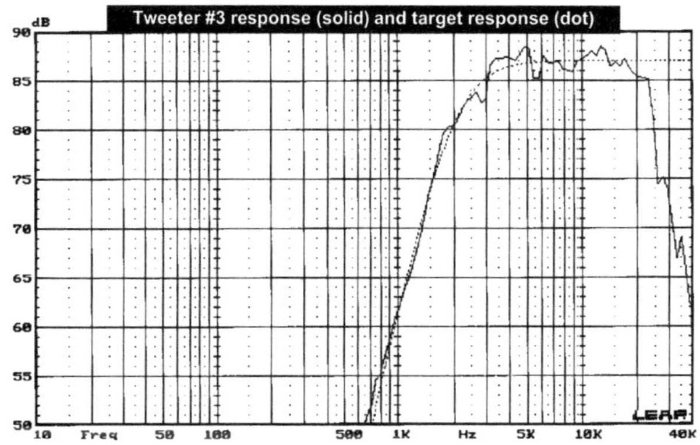

— Tw6 p17
······· Wofgeneric

FIGURE 2.56

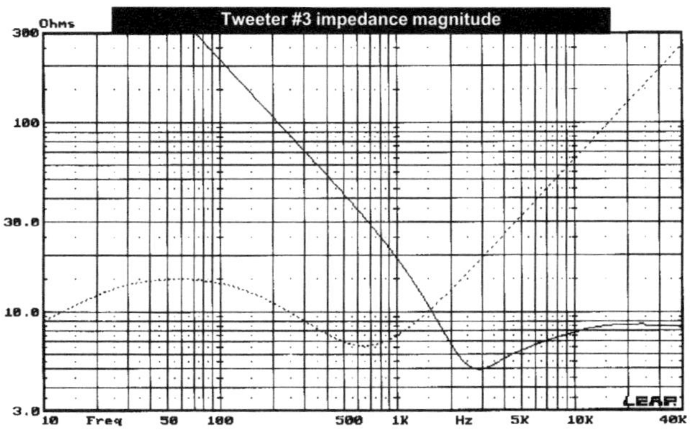

— Tw6 p17
······· Wofgeneric

FIGURE 2.57

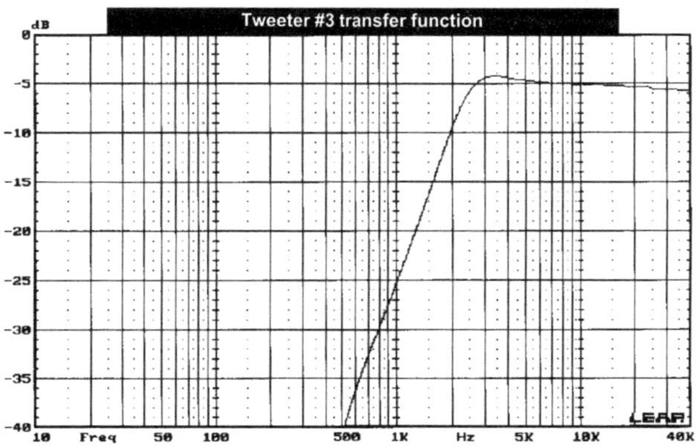

— Tw6 p17
······· Wofgeneric

FIGURE 2.58

2.69, the combination of woofer #1/tweeters #1 and #3 both tilt the upper end somewhat and produce response variations that exceed by a marginal amount the ±2dB goal. For this reason, my final choice was LP woofer #1 and HP tweeter #2.

For the on-axis and 30° off-axis simulation curves for this combination, along with the simulated impedance and transfer functions, see *Figs. 2.70-2.73*. The individual response curves of the AC8 and D25AG both with and without the final network are compared in *Figs. 2.74-2.75*. The final tweeter circuit combined the same switchable boost and cut circuits as the first speaker in *Chapter 1*, as illustrated by the diagram in *Fig. 2.76*. The final crossover schematic for this speaker is depicted in *Fig. 2.77*, with values given in *Table 2.1*.

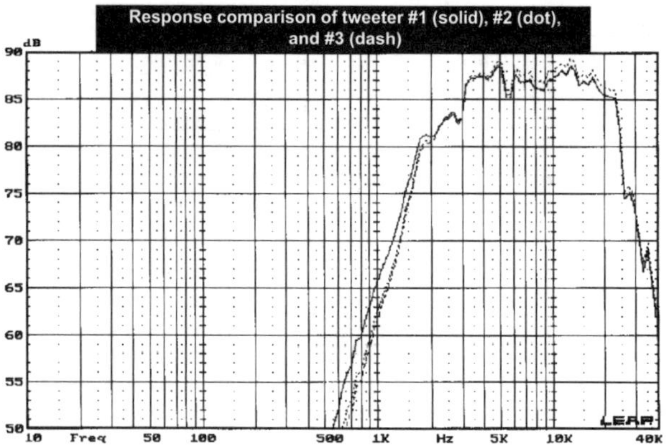

— Tw1
······· Tw2 p12
- - - Tw6 p17

FIGURE 2.59

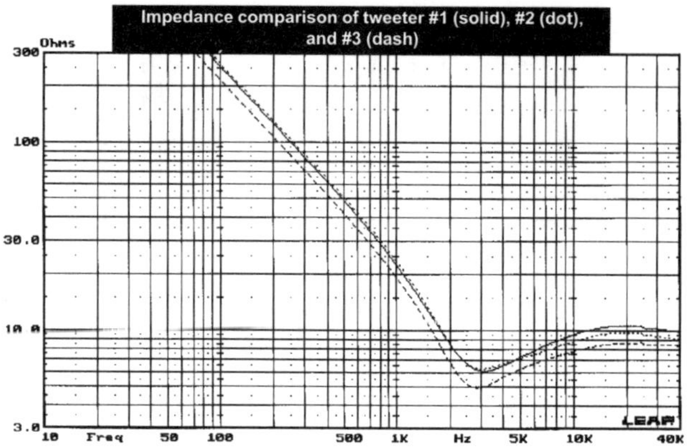

— Tw1
······· Tw2 p12
- - - Tw6 p17

FIGURE 2.60

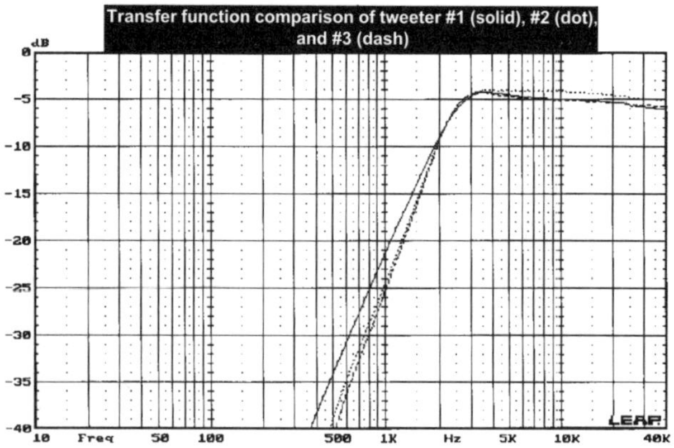

— Tw1
······· Tw2 p12
- - - Tw6 p17

FIGURE 2.61

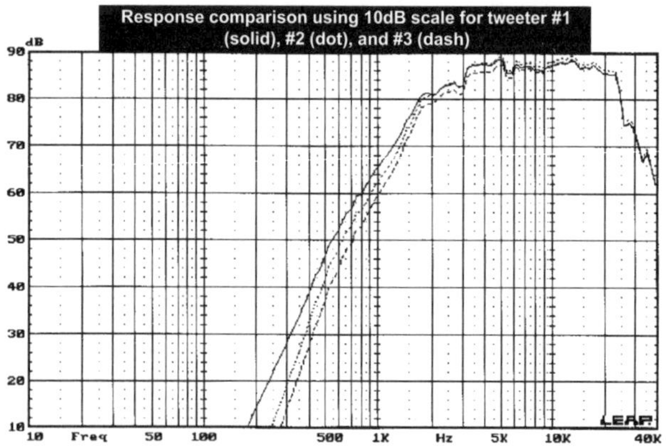

— Tw1
······· Tw2 p12
- - - Tw6 p17

FIGURE 2.62

TABLE 2.1
Crossover Component Values

Low-pass section

	Value	**Source**
C1	8.2µF, NP electrolytic	IT* (#1341048)
C2	100µF, NP electrolytic	IT (#1341085)
C3	6.8µF, NP electrolytic	IT (#1341045)
L1	1mH, 0.17Ω air core	IT (#0924)
L2	0.68mH, 0.26Ω air core	IT (#0056)
R1	5Ω, 10W	
R2	12Ω, 10W	
R3	6Ω, 10W	

High-pass section

	Value	**Source**
C4	6.1µF (3.9µF+2.2µF) polypropylene	IT(#1341335, 1341320)
C5	20µF (2 10µF) polypropylene	IT (#1341380)
C6	2.2µF polypropylene	IT (#1341320)
C7	1µF polypropylene	IT (#1341300)
L3	0.39mH, 0.18Ω air core	IT (#0053)
R3	4W (2 8Ω 3W metal film)	
S1	on-off-on DPDT switch	

*IT=InterTechnik

NOTE: The capacitors used in the low-pass section are all high-quality, smooth-film electrolytics, and do not require the typical bypassing. However, if you wish to do so, adding poly caps with values up to an additional 1µF will not affect the circuit's normal operation.

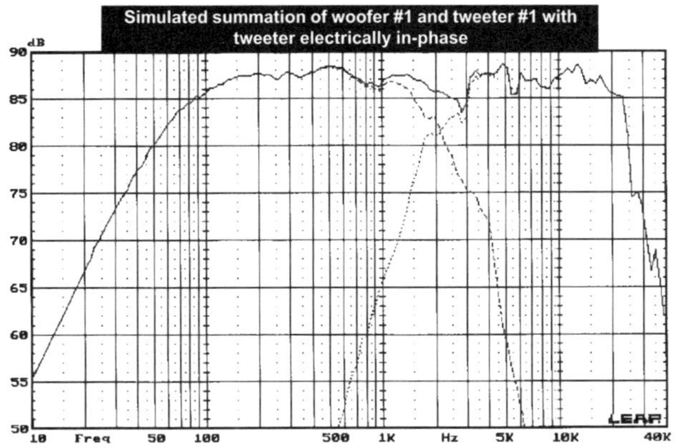

— 411
······· Tw1
- - - Wof 2 2nd + conj 4th at 2400 p10

FIGURE 2.63

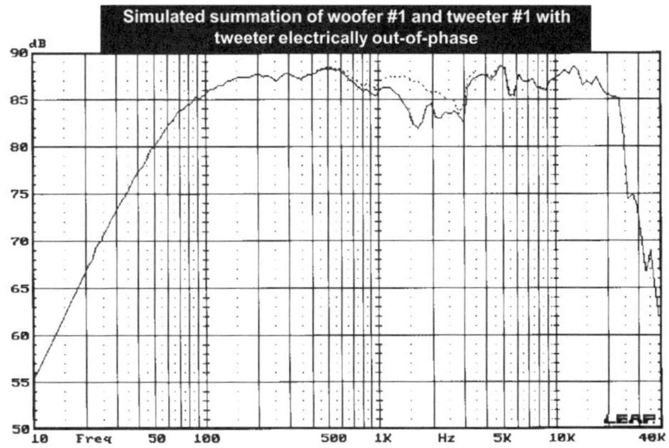

— 411R
······· 411

FIGURE 2.64

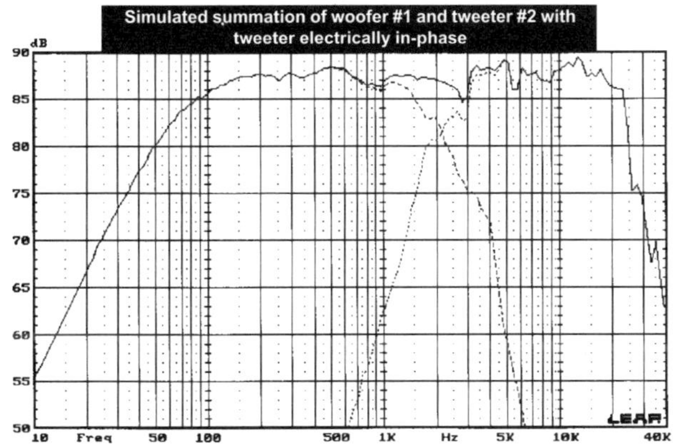

— 412
······· Tw2 p12
- - - Wof 2 2nd + conj 4th at 2400 p10

FIGURE 2.65

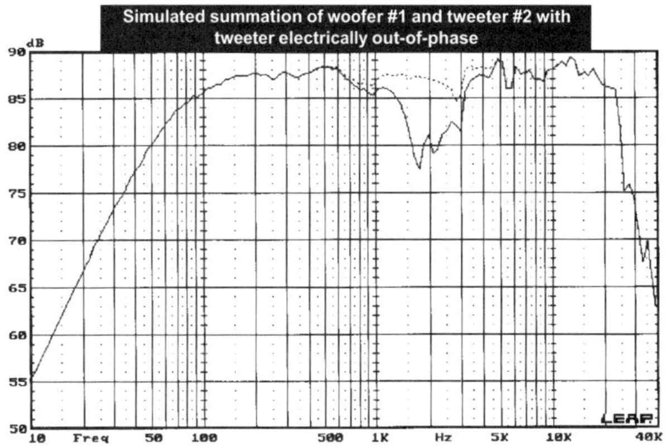

— 412r
······· 412

FIGURE 2.66

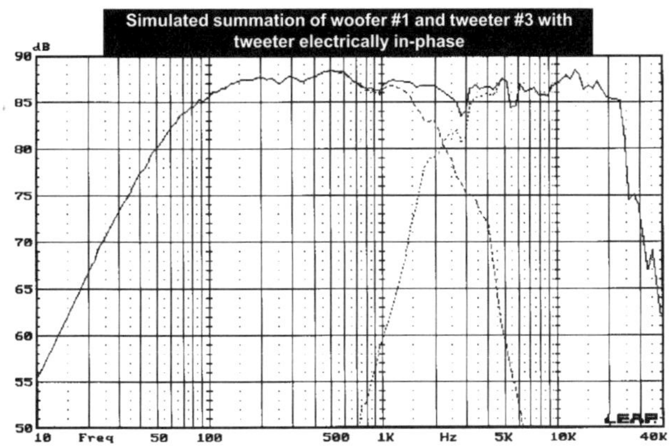

— 416
······· Tw6 p17
- - - Wof 2 2nd + conj 4th at 2400 p10

FIGURE 2.67

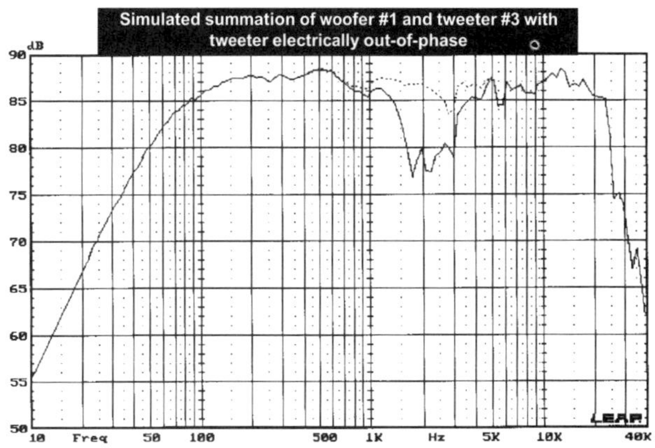

— 416r
······· 416

FIGURE 2.68

2.80 Measured Performance

Figure 2.78 depicts the prototype speaker's full-range on-axis frequency response, measured with the speaker tilted by 6° and the cut filter (CR conjugate) switch "in." The speaker's response is ±1.87dB from 76Hz-24kHz, and ±3dB from 58Hz-

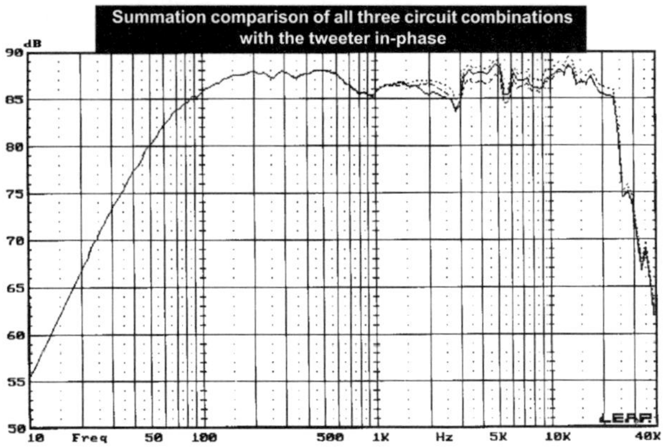

— 1011 --- 1016 ········ 1012

FIGURE 2.69

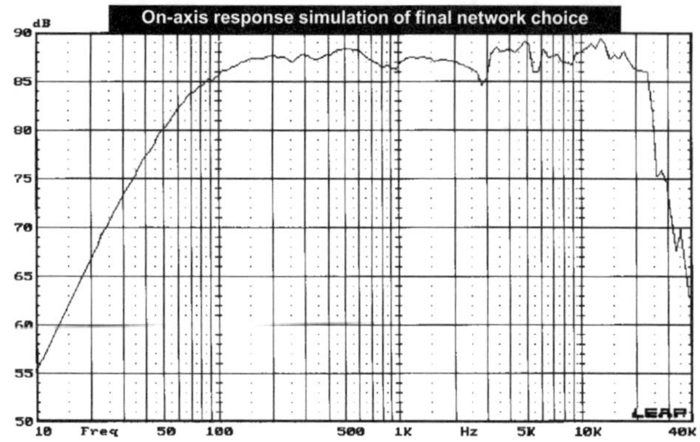

Wof 2 2nd + conj 4th at 2400 p10
Tw2 p12

FIGURE 2.70

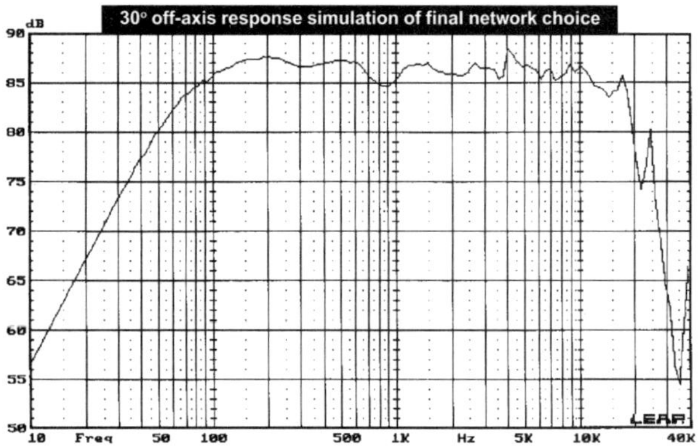

— Wof 2 2nd + conj 4th at 2400 p10
········ Tw2 p12

FIGURE 2.71

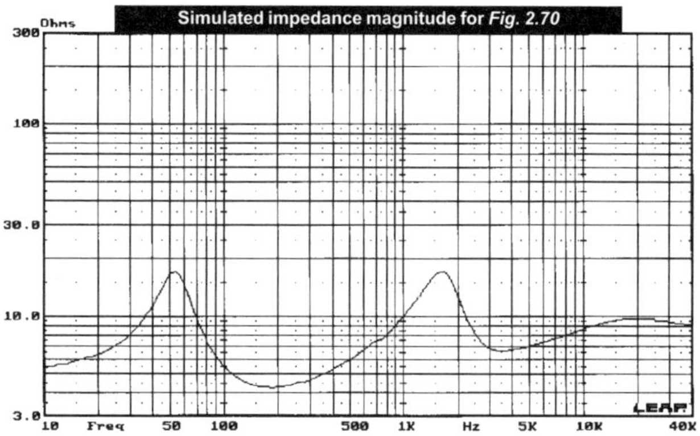

— Tw2 p12
········ Wof 2 2nd + conj 4th 2400 p10

FIGURE 2.72

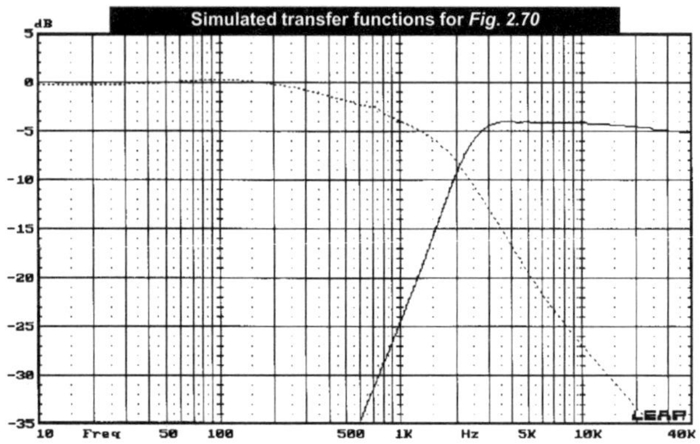

— Tw2 p12
········ Wof 2 2nd + conj 4th at 2400 p10

FIGURE 2.73

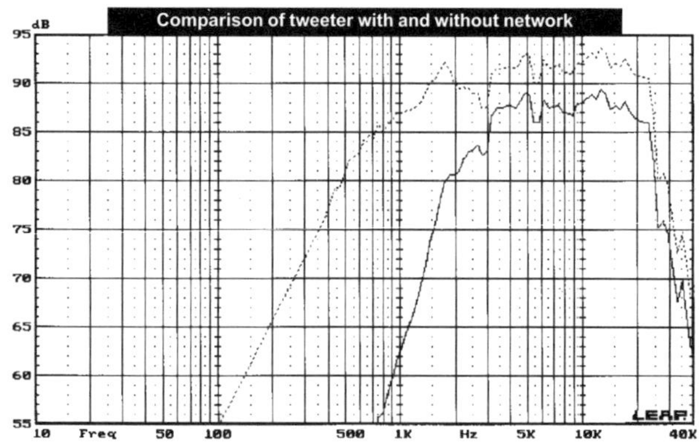

— Tw2 p12
········ Vifa D25AG no xover

FIGURE 2.74

25kHz. This easily fits into the design criteria set in the introduction, as well as being very close to the simulation in *Fig. 2.70* and well within 1dB.

As was expected, the off-axis response (illustrated by the 0°, 15°, and 30° curves in *Fig. 2.79*) is very good. The speaker is still ±2.5dB at 30°. The off-axis profile remains good out to 45°, as depicted in *Fig. 2.80*. The 45° off-axis curve is nearly as flat as *Chapter 1*'s 6.5" speaker, with only a small dip at 2.1kHz about 2dB lower than the other's. In addition, this speaker has an off-axis response out to 90° (*Fig. 2.81*), which is nearly as good as that of a 6.5" woofer.

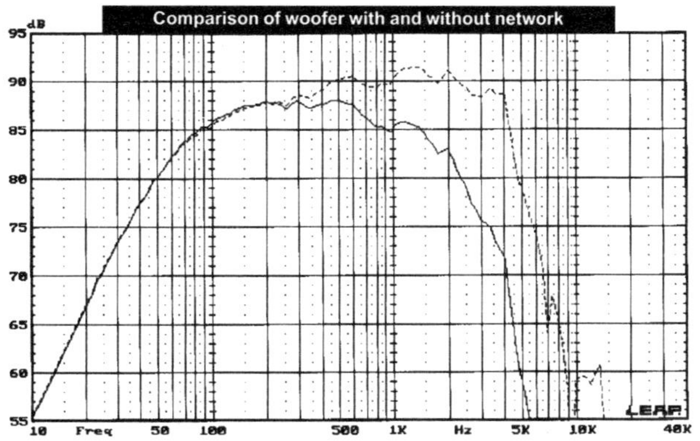

—— Wof 2 4th at 2400 p3
- - - AC8 wof no xover

FIGURE 2.75

Schematic of high-pass circuit with boost and cut networks

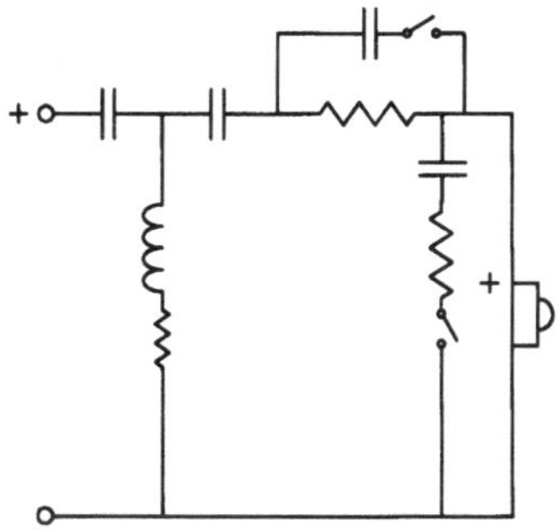

FIGURE 2.76

Final crossover schematic for AC8 and Vifa D25AG

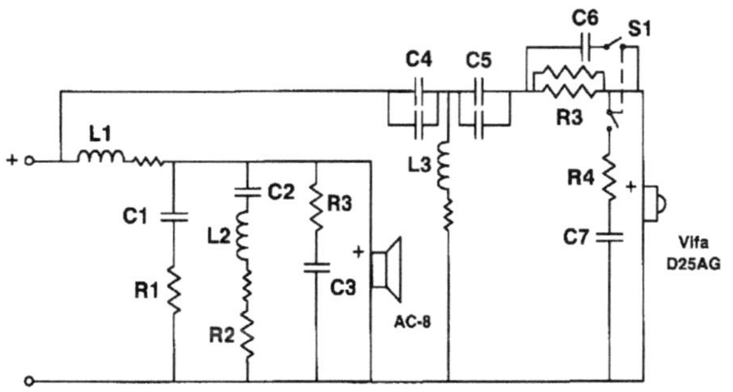

FIGURE 2.77

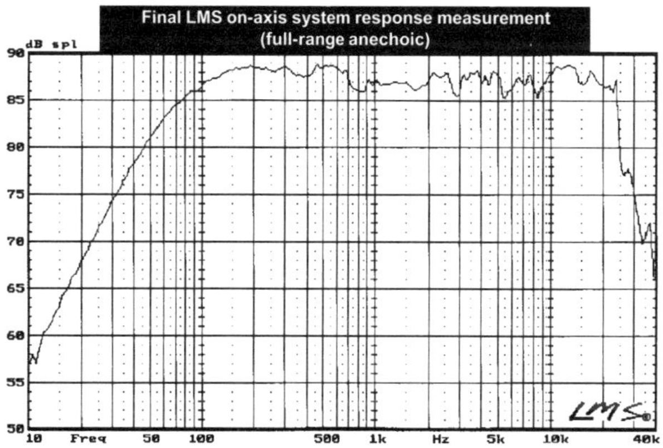

—— Fr -2dB

FIGURE 2.78

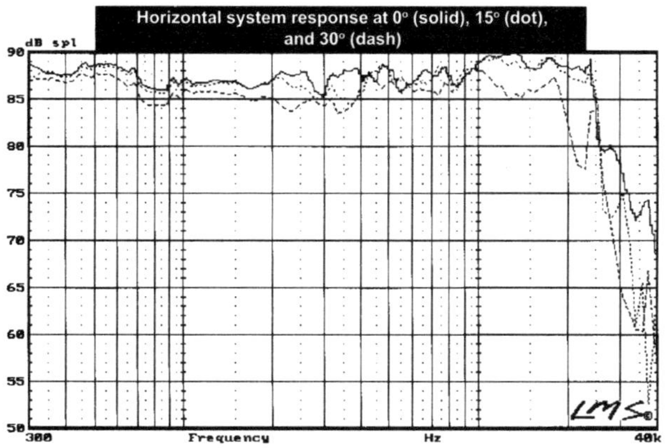

—— Full range
········ M4a 15
- - - M4a 30

FIGURE 2.79

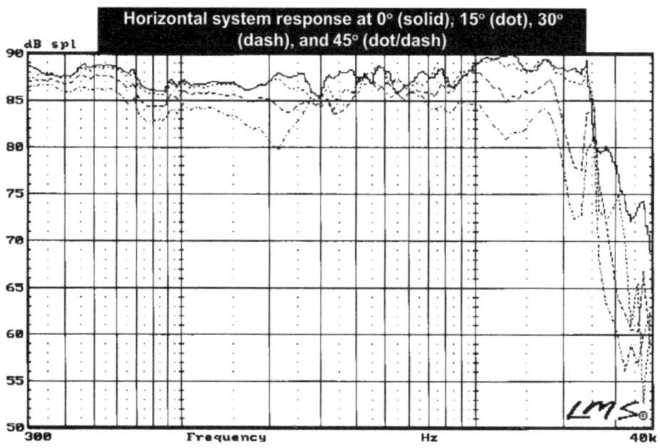

—— Full range
········ M4a 15
- - - M4a 30
-·-·- M4a 45

FIGURE 2.80

The vertical off-axis plots shown in *Figs. 2.82* (over the top of the cabinet) and *2.83* (under the bottom) exhibit the typical lobing caused by noncoincident driver mounting. When the 15° left/right and up/down curves are averaged with the on-axis curve, the response window remains at ±2.5dB. (The scalar RMS average curve of 0°, 15° left/right, and ±15° in *Fig. 2.84* would be ±2dB if the tweeter conjugate had been engaged.)

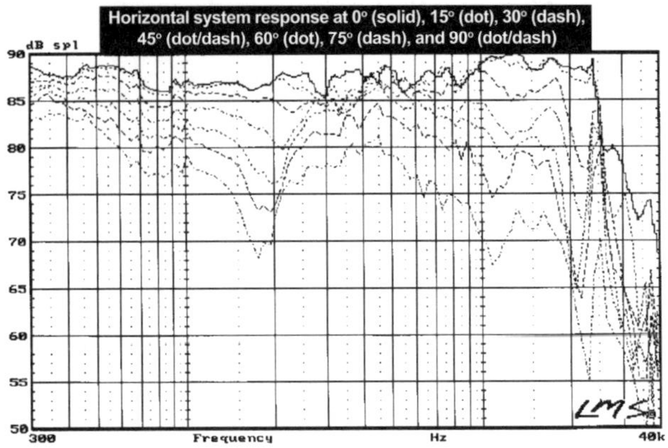

— Full range
······· M4a 15
--- M4a 30
-·-· M4a 45
······· M4a 60
--- M4a g75
-·-· M4a 90

FIGURE 2.81

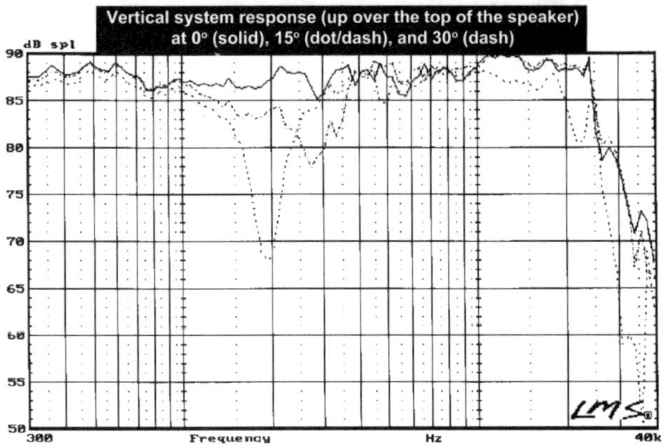

— M4a 0 h
-·-· M4a +15
- - - M4a +30

FIGURE 2.82

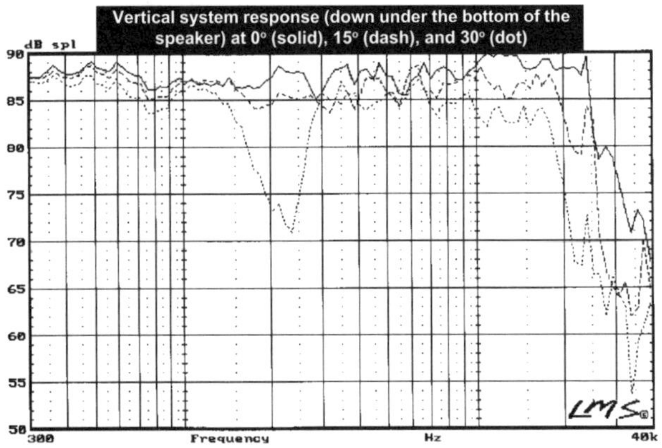

— M4a 0 h
--- M4a -15
······· M4a -30

FIGURE 2.83

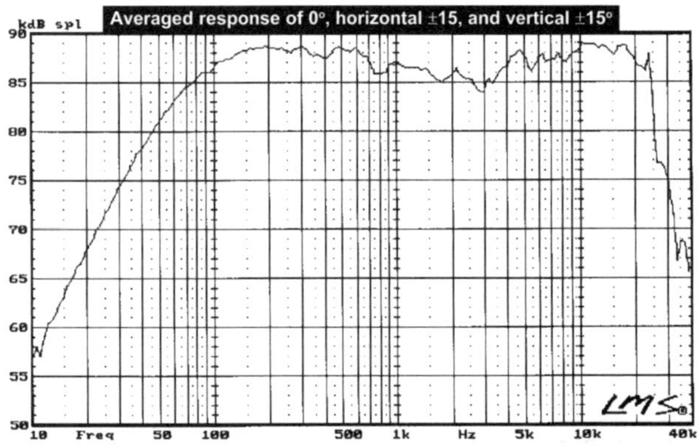

— Average

FIGURE 2.84

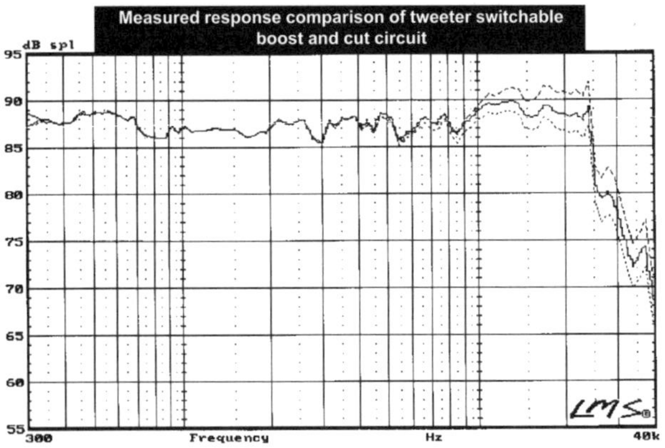

— Full range
······· M4a 0 tw conj1
--- M4a 0 tw conj2

FIGURE 2.85

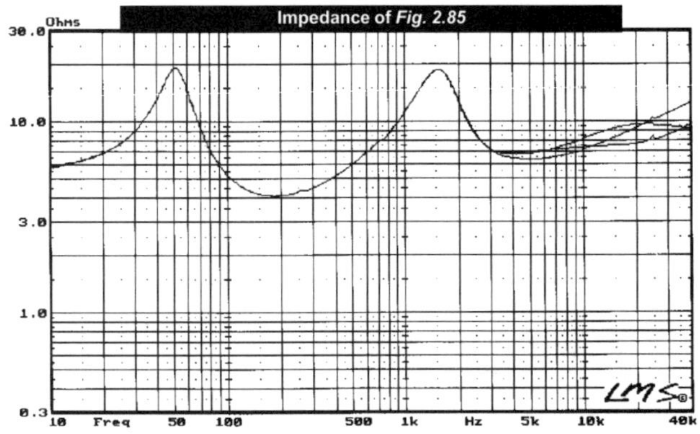

— M2 sys imp
--- M2 sys imp tw -
-·-· M2 sys imp tw +

FIGURE 2.86

The tweeter contour network provides a 2dB boost and 1dB cut, as depicted in *Fig. 2.85*, with the effect on the impedance seen in *Fig. 2.86*. In terms of matching left and right channels, the deviation for the two prototypes was no greater than 1dB and typically 0.5dB, as illustrated in *Fig.*

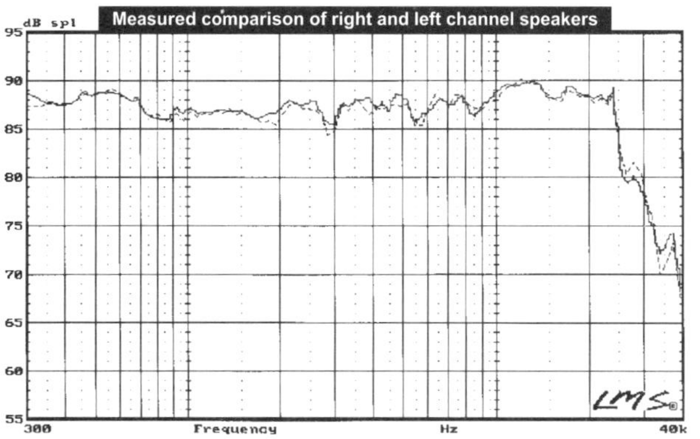

— Full range
--- M4b 0

FIGURE 2.87

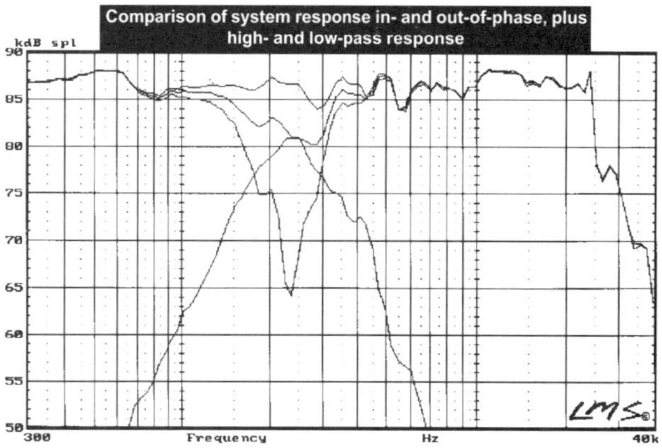

— Sys 0 3ohm
— Sys 0 rev
— Wof 0
— Tw 0

FIGURE 2.88

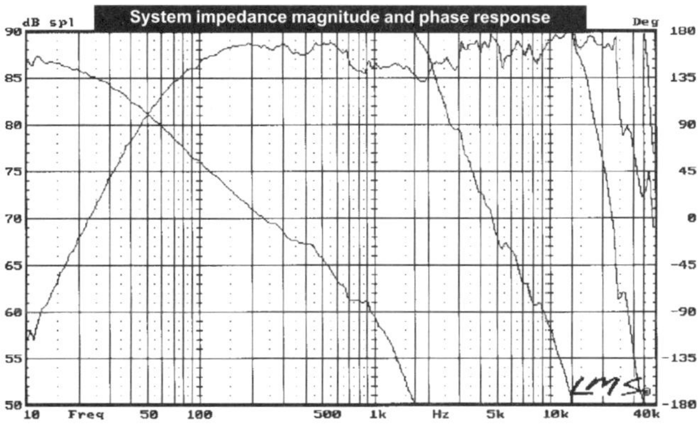

— 100uS

FIGURE 2.89

Complex impedance curve (Nyquist plot)

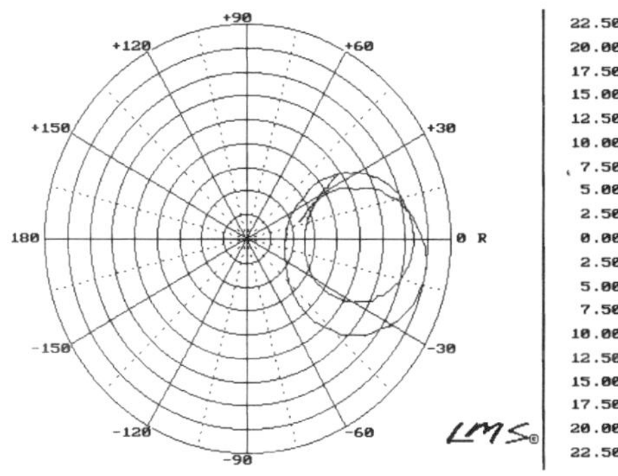

-·-· M2 sys imp

FIGURE 2.90

2.87. I achieved this close matching using only two samples of each driver, although this may not always be the case.

Figure 2.88 gives the in-phase system response compared to the out-of-phase response, as well as the individual high and low-pass responses. Each section's acoustic slope is fairly symmetrical, with reasonably deep null close to the crossover frequency (2.5 and 2.4kHz). The null is not completely symmetrical, however—the upper frequency end is steeper even though the response is additive throughout the crossover region—indicating that the drivers are close but not exactly in-phase. While not really that bad, this condition emphasizes the difficulty in achieving a completely symmetrical transition with an excess of 50μs of physical delay between the two drivers.

The speaker's calculated acoustic phase response is given in *Fig. 2.89*. I achieved it using the same technique of mathematically adding delay to the low-pass response and summing it to the high-pass response in the LMS software.

The complex impedance curve (Nyquist plot) is illustrated in *Fig. 2.90*. It shows a maximum inductive phase angle of 37.8° at 867Hz and a maximum capacitive phase angle of 53.5° at 73Hz, neither of which is considered excessive in terms of amplifier operation. The impedance minimum of 4Ω at 190Hz (*Fig. 2.86*) precludes paralleling two of these speaker into one channel, however, unless the amplifier is adept at driving loads as low as 2Ω.

The MLSSA cumulative spectral decay plot (*Fig. 2.91*) is uncluttered, with the only sharp resonance indicated at 25kHz, the break-up mode for the aluminum dome tweeter. By 2ms, upper harmonics (above 2kHz) have settled below -30dB, which is normal unless the woofer has excessive break-up resonances or the crossover has a particularly undamped section.

MLSSA cumulative spectral decay plot

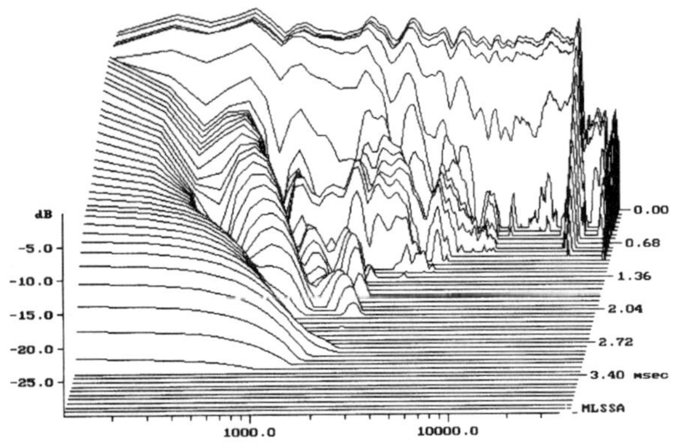

FIGURE 2.91

MLSSA accelerometer CSD plot of box side panel

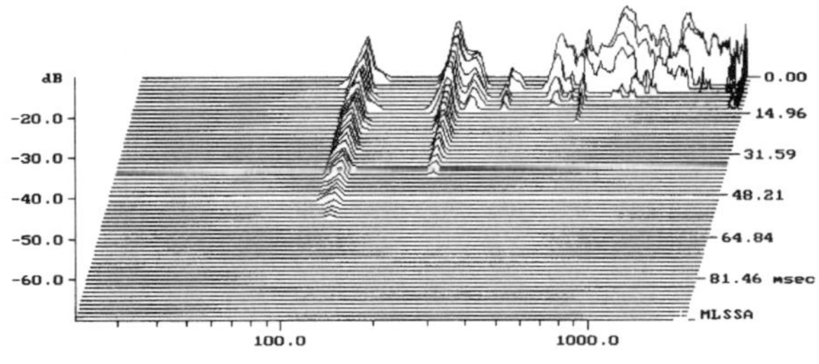

FIGURE 2.92

MLSSA accelerometer CSD plot of box top panel

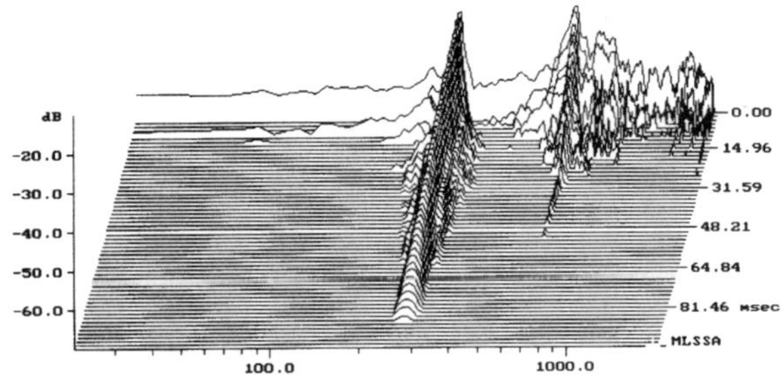

FIGURE 2.93

MLSSA accelerometer CSD plot of box rear panel

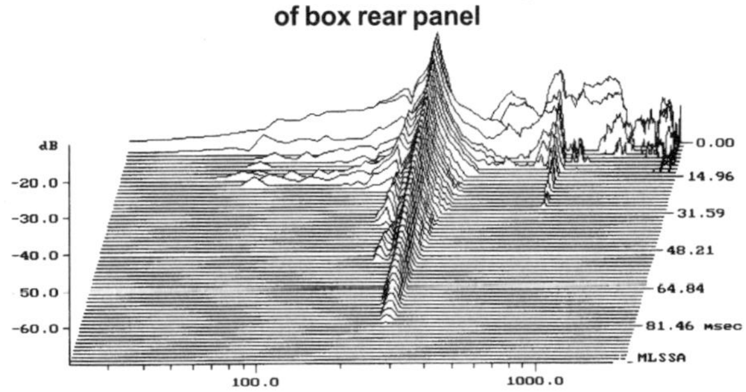

FIGURE 2.94

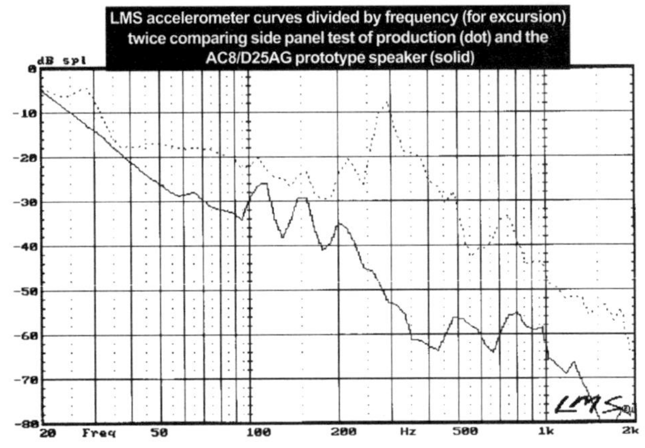

—— M2side/f/f
· · · Mbside/f/f

FIGURE 2.95

2.90 Cabinet Performance

This speaker's construction is identical to the Vifa P17WJ/Scan-Speak D2905 in *Chapter 1*, and likewise uses an internal Sonotube semicylindrical baffle backfilled with sand. This technique creates a very substantial enclosure: the AC8/D25AG prototype weighed 50 lbs. without crossovers and stands. I performed accelerometer tests in the same manner as previously described, using the piezo-element transducer in conjunction with MLSSA (*Figs. 2.92-2.94*) and LMS. The MLSSA CSD plots can be compared to the production loudspeaker plots in *Figs. 1.118-1.120*. Obviously, this cabinet is also well damped.

I converted the accelerometer curves for the box side walls to read more like excursion curves by dividing the curve by frequency twice. The results are compared to the side wall reading on the production loudspeaker in *Fig. 2.95*. As expected, the lower-frequency amplitude is greater than the box resonances at higher frequencies.

2.100 Construction Details

The cabinet construction details, diagramed in *Fig. 2.96*, are for the most part identical to those of *Chapter 1*'s speaker. The damping characteristics of this construction are quite good, even though the techniques are relatively simple. I cut the top, bottom, rear baffle, and left and right walls from 1" MDF, and assembled them using glue and staples (nails). The four walls are butt jointed rather than mitered for ease of assembly.

I limited the internal reflections and standing-wave modes using an internal rear enclosure wall made from 10-inch-diameter cardboard tubing (Sonotube). Slicing the 0.25-inch-thick cardboard in half lengthwise, I formed a semicylinder and cut it to the inside height of the box. After I had positioned it to the rear of the enclosure and stapled it to the back and side walls, I sealed the edges of the cardboard with a combination of hot glue and silicone. This created two empty spaces in the corners of the box.

Once the cabinet was completely assembled, wired, and finished, I drilled 0.5" holes in the bottom of the enclosure in the center of the corner spaces. I then poured sand (the fine white silicon variety sold for gardening) through the holes into the area behind the cardboard tubing, filling it completely. I then sealed the holes with a dowel plug. If you are using a terminal cup, as I did with the prototype, it must be located to the side of the enclosure, due to the location of the semicylindrical cardboard.

Bracing consisted of two 1.125" hardwood dowels. I placed one exactly between the woofer and tweeter, extending through the cardboard tube from the front to the rear baffle.(I cut the hole into the tube before it was mounted.) The second dowel brace is positioned slightly below the first brace from side wall to side wall. (In the box construction diagram, the brace locations are noted by small black circles.) Since cutting the dowels exactly to length can be a problem depending upon your woodworking skills, I predrilled a shallow 0.25" countersink into each wall to position them and allow them to be cut " longer than the enclosure dimension.

I applied construction adhesive to the dowel insets and set the dowels into position during the box assembly process. Once the adhesive sets, the bracing is secured to the enclosure. This procedure avoids the problem of dowels which have been cut too long bowing the enclosure out, or, if cut too short, causing the walls to suck inward (especially if nailed or stapled through the wall).

The front baffle, 1.75" in depth, is made from two layers of MDF: a 1" bottom and a 0.75" top. To assemble it, I first glued and stapled the bottom layer to the box, and then attached the top layer to it in the same manner. Routing this top baffle section countersunk the drivers flush with the baffle surface. To prevent the woofer from acoustically loading within the 1.75" channel, I routed the 1" baffle layer with a curved 45° bit on the inside of the woofer hole opening into the enclosure.

I made the 0.75" baffle layer with a 0.75" inset on both the sides and top to accommodate the 0.75"

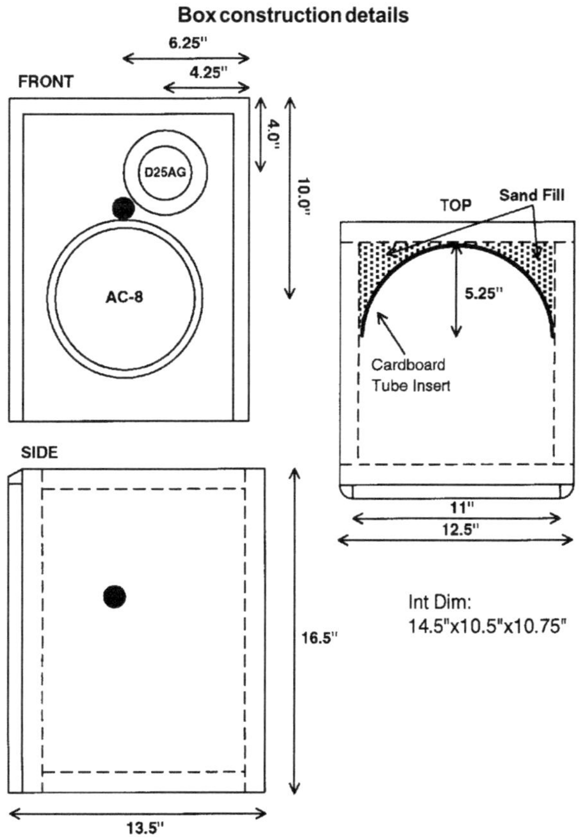

FIGURE 2.96

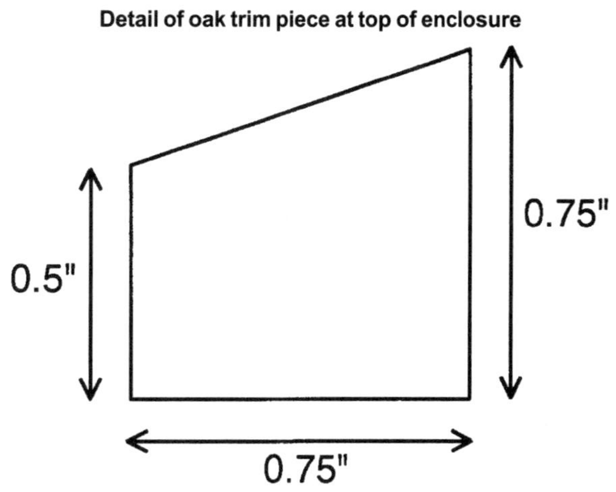

FIGURE 2.97

oak quarter-round trim on the baffle sides and the angled trim piece at the top. (Be certain to measure the dimension of the oak quarter-round carefully because it can have considerable variance.) Details of the oak top trim piece, which extends across the top of the front baffle's 0.75" layer, are illustrated in *Fig. 2.97*. The oak quarter-round is cut at the

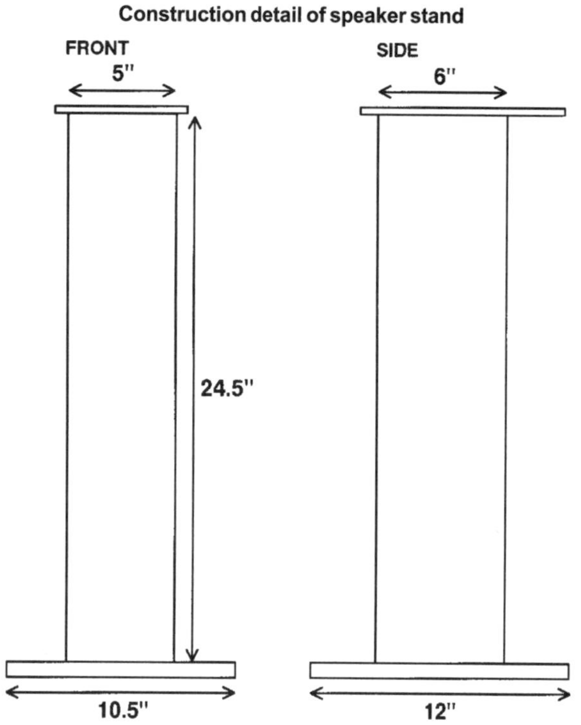

FIGURE 2.98

same angle as the top trim piece for a smooth angle across the top of the enclosure front.

Once the cabinet was assembled, I painted the front baffle and enclosure bottom, contact cemented flat black laminate to the sides, top, and rear, and glued the oak trim in place. Internal wall panels (sides, top, and bottom, excluding the cardboard tubing) were further damped with two layers of Antiphon Type A-13 extensional damping material. This 1/16-inch-thick, black bituminous felt and clay material comes with self-adhhesive backing. I covered approximately 50-60% of each wall panel area. An alternative material (unavailable during construction of the prototype) which is as effective, if not more so, is the Black Hole Pad sold by Orca Design & Manufacturing. A single layer of this thicker material is sufficient.

For damping standing waves within the enclosure, I used AC Foam (an egg-crate type of acoustic foam provided by ACI) to line the four walls and rear. The area immediately behind the woofer, which accounts for about one-half the entire volume, I filled—not stuffed—with slightly teased Acousta-Stuf. If the density is increased, the Q will be lowered. Since this speaker already has a rather low Q_{TC} (measured at 0.65), excessive stuffing is inappropriate.

I mounted the woofer and tweeter using Well-Nut inserts. Sold by Pop Fastener Corp., these rubber inserts have a nut mounted in the end and are frequently used as a vibration damping mount for electric motors. They mount from the top of the hole, allowing easy removal of the driver and providing additional speaker isolation from the cabinet. In conjunction with the Well-Nuts, I used a foam tape gasket, the same type that is used for weatherstripping. A better neoprene rubber gasket tape is also available from North Creek Music Systems.

Unlike with most speakers, the crossovers were not located inside the enclosure, but placed inside the stand's vertical stem portion. This not only isolates the components from vibration and the influence of the drivers' magnetic fields, but also allows for a much larger component layout and facilitates placing the inductors at greater distances than normally would be possible.

Two terminal sets—one for the woofer and one for the tweeter—are thus required on the rear of the enclosure. For this purpose, I used a round terminal cup provided by ACI. The small cup, which fits a 2" hole, is less prone to vibration than some of the larger ones (which are almost like small passive radiators), and is fitted with a pair of gold binding posts. To these, I added an additional pair of identical gold binding posts from Madisound.

The four terminals provide direct connection to both woofer and tweeter, with the wiring from the terminal routed through a hole in the Sonotube and sealed airtight with silicone. Since this is a sealed box, a small bead of silicone should be applied to all internal joints to ensure the enclosure's air-tight integrity.

I fabricated grille frames from 0.25" MDF, with 0.25" standoffs for the snap-type plastic grille fasteners. The standoffs are 0.5" x 0.5" pieces of 0.25" MDF glued to the frame. They provide an attractive cosmetic look, as well as the thickness required to mount the plastic fastener. I attached black doubleknit cloth to the frame using 3M spray fabric adhesive. This method works well, although you may easily alter this design aspect.

Construction details for the stands are shown in *Fig. 2.98*. I used 0.75" MDF for both the base and the vertical stem, and finished the base in black laminate with oak iron-on tape on the edges. To mount the brass cone-type feet to the bottom of the base, I first drilled shallow countersunk holes. I attached the base to the vertical stem, also finished with black laminate, using 1.5" black wood screws. To further reduce vibration transferred from the speaker cabinet, I covered the inside walls of the stem with Black Hole Pad.

A rectangular blank terminal cup served to mount the three pairs of gold binding posts: one for the input; one for the crossover's high-pass section to connect the tweeter; and one for the crossover's low-pass section to connect the woofer. In addition, I mounted the tweeter high-frequency contour switch on the cup panel. Another pair of terminals could easily be added to provide biwire capability if desired. Banana plug jumpers connect the crossover in the base to the drivers in the speaker enclosure.

Details of the stand top are depicted in *Fig. 2.99*. The screw-in-type spike, which uses a brass insert (provided by Meniscus), can be adjusted until the front of the speaker is elevated to produce the required 6° tilt. Both the spike and the rubber pad help isolate the speaker from the stand, while the corner bracket at the rear of the top piece secures them. I mounted the top piece to the stand stem in the same manner as the base using four wood screws, but I also used rope caulk (the kind used to seal windows and doors) as a gasket between the top and the vertical stem to provide further isolation. Because its depth is several inches greater than *Chapter 1's* prototype, this speaker's stand top is placed slightly off center.

The crossover is mounted on 0.25" hardboard cut to fit the interior side wall of the stand stem. High- and low-pass sections were produced on separate boards, although they could also be placed on a single mounting board. The idea is to use the stand's entire vertical height to spread out the components in order to provide maximum separation of the inductors. Due to this mounting configuration (i.e., the inductors placed some distance apart), I did not take the normal precaution of placing adjacent inductors at right angles and perpendicular to each other.

Even though they were mounted in the same plane, inductive coupling did not appear to be a problem when I measured individual sections to validate my assumption. You can provide about 10-20dB more attenuation, however, by mounting them perpendicular at 90° angles if you wish to use some kind of strapping to hold the inductors in place.

Component layout is not critical except for the placement of the inductors, as shown in the photograph in *Fig. 2.100*. I mounted the woofer section inside the stem at the base of the stand, using silicone adhesive to affix it to the side wall. The tweeter section was mounted in the same manner, with the inductor at the top of the vertical stem. For internal wiring in both the stand and the speaker cabinet, I used Monster cable for the woofer section and Monster XP for the tweeter section.

I point-to-point wired the crossover components using brass split rivets as soldered connection points. This also provides a physical mount to the board for capacitors and resistors. To use the split rivets as connection points, lay out the components on the board, locate junction points and mark them with a pencil, then drill the board and insert the rivets. I attached inductors to the crossover boards with silicone adhesive.

Construction detail of speaker stand top

Top Dim: 0.5"x6"x9.5" MDF

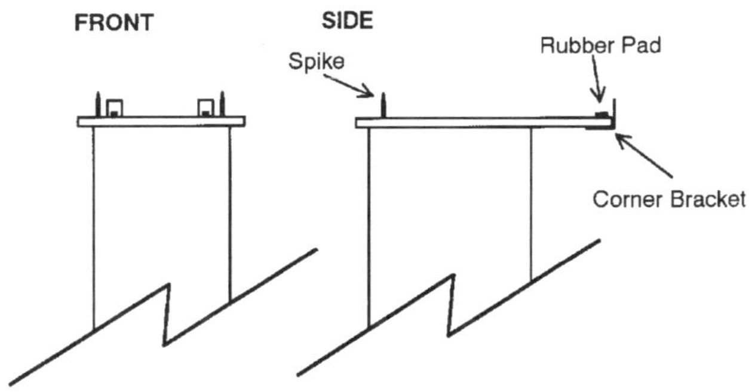

FIGURE 2.99

Photograph of crossover layout

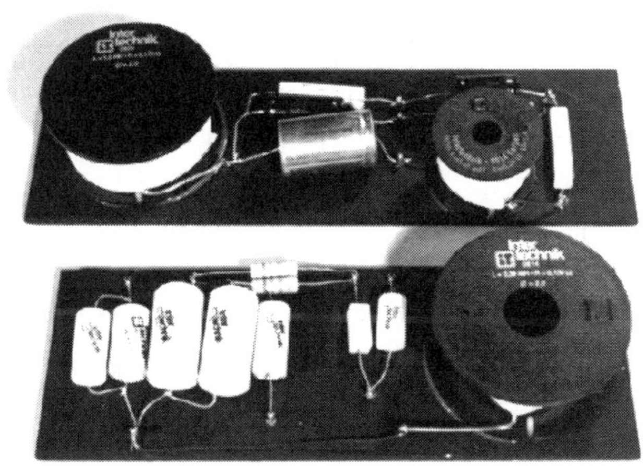

FIGURE 2.100

2.110 Adding a Subwoofer

As with *Chapter 1's* prototype, adding a subwoofer with parameters optimized for low-bass performance can enhance any small two-way speaker and provides in my opinion one of the best overall speaker formats available. Even though this unit's low-frequency performance is satisfactory (depending greatly on your room and the speaker's location in it), the addition of the 8" subwoofer described in *Chapter 3* definitely adds to the low-bass impact and depth.

The active crossover I used for this project is the same TDM Model 24CX-2 (depicted in the

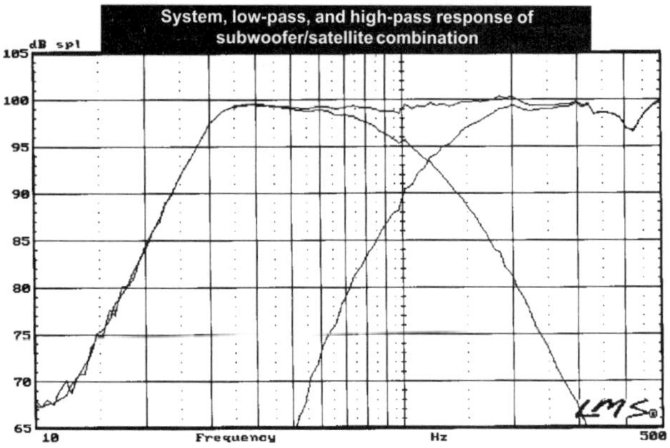

-·-· M2
- - - M2 sub
—— M2 sat

FIGURE 2.101

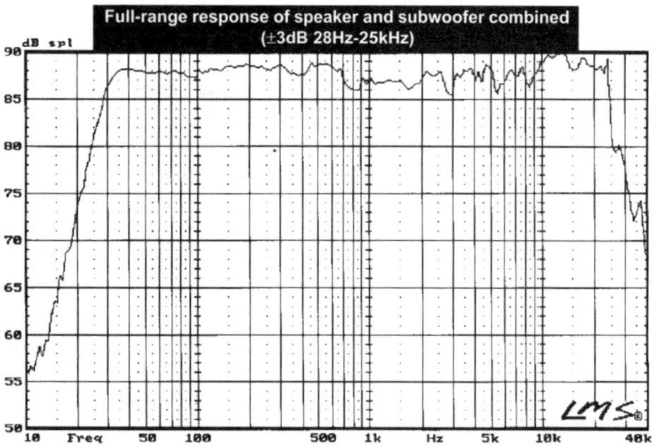

- - - M2 + sub

FIGURE 2.102

photograph in *Fig. 1.127*), which provides 24dB/octave L-R slopes with variable frequency controls. *Figure 2.101* shows the groundplane measurements of the 8" subwoofer and the two-way 8" prototype speaker. *Figure 2.102* gives the full-range response of the two-way speaker and the subwoofer, showing a very extended and reasonably flat bandwidth of ±3dB from 28Hz-25kHz.

Chapter Three

A TWO-WAY SATELLITE WITH SUBWOOFER

3.10 Overview

Our third two-way speaker design represents a radical departure from our first two. Although the drivers used are a 7" woofer and a 1" dome tweeter, which represents nothing unusual, the application is entirely different. This design was conceived as a biamped three-way speaker, incorporating a passive two-way crossover in a two-way satellite, with a separate subwoofer and active crossover. However, our real departure comes in the enclosure design for the two-way satellite. Rather than optimize the woofer box design for maximum low-end performance, the enclosure is designed to be part of the subwoofer electronic crossover.

I will explain this in more detail, but the idea of using the woofer box design as part of the active electronic crossover consists of first locating a woofer which will produce a critically damped closed-box response Q of 0.7 in a small sealed box and also have a low-frequency rolloff (-3dB frequency) of 100-120Hz. The second-order Butterworth woofer response, which is a high-pass filter, is then cascaded or combined with a second-order Butterworth active filter high-pass response to produce a fourth-order Linkwitz-Riley (L-R) high-pass. We could use the same basic approach to create a passive fourth-order L-R filter response, which we can accomplish by placing two second-order Butterworth (Q = 0.707) filters in series (cascaded) with each other.

With this speaker, the woofer enclosure provides one second-order section and the electronic crossover provides the other second-order section. Either way the resulting response is the same fourth-order L-R response with a Q of 0.499. Don't let anyone convince you that fourth-order filters are not the combined Qs of two second-order filters.

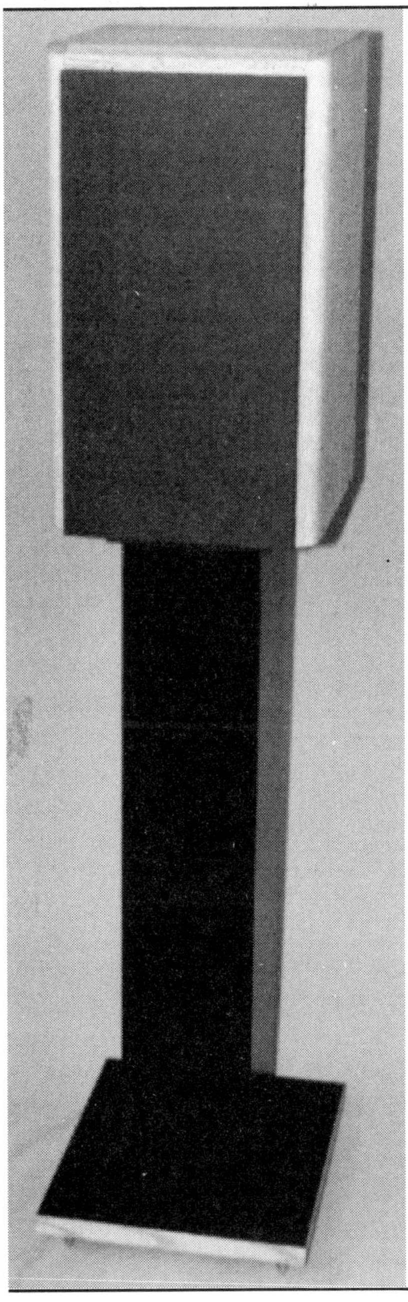

From this uncommon technique we achieve two unusual results. The first is the electronic crossover. Most of the home audio types have symmetrical crossover slopes, i.e. both the high- and low-pass crossover sections have the same slope and filter Q. Even if they allow us to vary the frequency, they generally don't allow us to "mix" slopes. What we need for this subwoofer/satellite combination is an electronic two-way crossover which has a fourth-order L-R low-pass, and a second-order Butterworth high-pass. Fortunately, Marchand Electronics, who produce a very respectable line of home audio electronic crossovers, have such a unit (as do other "kit" crossover suppliers). The Marchand XM16 can be readily configured for a variety of high-pass and low-pass slopes and is an ideal choice for this project.

The second unusual design feature is a woofer enclosure which rolls off in the 100-120Hz region.

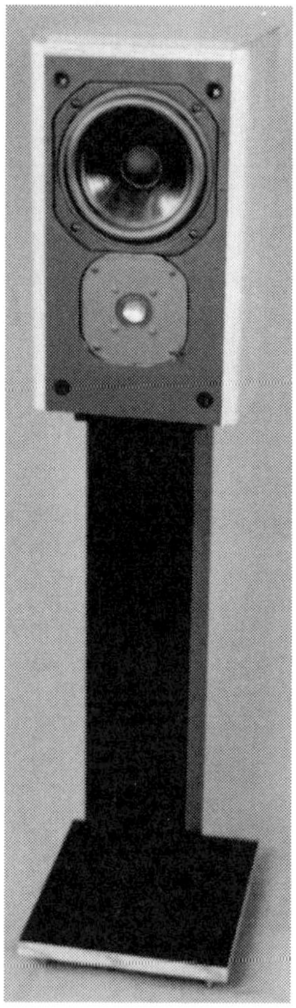

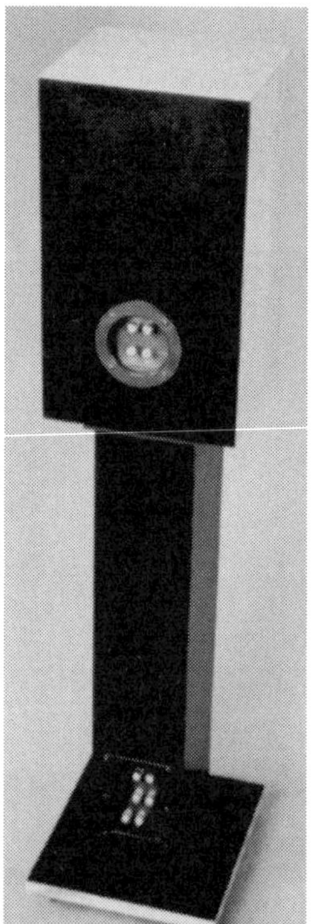

A by-product of the unique crossover/enclosure system configuration, which requires a sealed-box enclosure with a high roll-off frequency, is a low-excursion and high-power handling midbass. The "magic bullet" turned out to be the Focal 7N515. For the high-frequency part of the satellite speaker I used the Focal T120ti titanium inverse dome. The subwoofer driver is Vifa's M22WR-09-06, a spectacular 8" woofer.

3.20 The Woofer

As with all the drivers in this book's projects, the Focal 7N515 is a high quality design. The 7N515 has a cast metal basket, a curvilinear Neoflex cone (a proprietary isotopic synthetic cone material), a neoprene rubber surround, and a 40mm Kapton voice coil former with a high power handling single layer, edge-wound copper ribbon voice coil. Focal rates this woofer at 125W nominal power handling and a sensitivity of 89dB 1W/1m.

After the break-in period, I tested the samples and calculated the T/S parameters as follows:

	A	B
Z_{NOM}	8	8
R_E (Ω)	7.66	7.66
S_d (M^2)	0.0143	0.0143
Bl (TM)	11.45	11.28
V_{AS} (ltr)	34.3	34.3
C_{MS} (μM/N)	1180	1180
f_S (Hz)	33.2	33.2
Q_{MS}	2.717	2.664
Q_{ES}	0.237	0.237
Q_{TS}	0.218	0.224
SPL (2.83V)	88.9dB	88.8dB
L_E (1kHz)	1.29mH	1.29mH
(20kHz)	0.47mH	0.48mH

The parameters for the two samples are about as close as you can get having two different drivers, so I'll forego comparing samples. Since this design calls for a specific target closed box Q_{TC} of 0.7, I don't need to consider the various enclosure alignments which would be suitable for the 7N515. Instead, let's go directly to an analysis of the box volume which achieved this target.

Using the LEAP Quick Cabinet Utility, a sealed Butterworth alignment for the 7N515 would require a box volume of only 0.12 $ft.^3$, which is not much larger than a good sized cereal bowl. *Figures 3.2-3.5* compare the 1W and 50W computer simulations (which included a 0.5Ω series resistance) for the driver's frequency response, group delay, cone excursion, and cone velocity respectively. Running the LEAP screen cursor down to the -3dB point on the 1W frequency response curve the frequency reads 106Hz with a phase angle of 89° (88° without the series 0.5Ω). Since a phase angle of 90° at the -3dB point indicates a Q of 0.7, the chosen alignment, the 0.12 $ft.^3$ volume simulation is quite close (the simulation also included a 50% fiberglass stuffing factor).

At 50W, the driver damping has typically decreased, and the SPL output has increased to 103-105dB. The cone excursion curves in *Fig. 3.4* shows it to be at a maximum of 3.4mm at 90Hz. Since the 16.5mm voice coil length and 7.7mm gap height give the woofer an X_{MAX} of 4.4mm, the 7N515 is comfortably operating within its linear operating range at 105dB SPL. This is before applying the electronic filter to the low-end driver response. Together, the biamped 7N515 with its high power handling capacity and its edgewound voice coil should be as nearly a bullet-proof midbass driver as anything I have ever seen.

I built the prototype using some of the same box construction procedures used with the first two speakers and incorporated a semicylindrical section of 8-inch diameter Sonotube as an interior rear baffle. The driver volume (magnet assembly and cone) is a significant percentage of our small enclosure, so the position of the tube had to be carefully calculated to yield to proper net volume. Once built, I completely filled this small volume with Acousta-Stuf. This material, unless overcompressed, does a good job of damping the high frequencies, and also doesn't lower the Q by significant amounts. I verified this by measuring the impedance with LMS (*Fig. 3.6*) and using LEAP to calculate the final cabinet Q_{TC}, which was 0.69, without the passive crossover. This is precisely as predicted by the computer simulation.

Figure 3.7 illustrates the initial outdoor groundplane measurement showing the -3dB point to be 120Hz, only slightly higher than our simulation, but still suitable as a crossover frequency for the biamplified subwoofer configuration. *Figure 3.8* shows the woofer's on- and off-axis response at 0°, 15°, 30° and 45°. As with the 6.5" Vifa woofer and the ACI 8" woofer, the 2-2.2kHz 45° off-axis curve is down about -3dB from the on-axis curve, but rapidly increasing to -10dB at 3kHz. The maximum crossover frequency which would ensure the absolute best off-axis performance is in the 2kHz region. The overall response is ±2dB from 300Hz-4.2kHz, which covers most of the range up to the expected crossover frequency.

Looking at the full-range anechoic response of the woofer in *Fig. 3.9*, it exhibits about the same rising response as do the woofers in *Chapters 1* and *2* when measured in their cabinets. The rise begins at 250Hz and increases by +4dB at 750Hz,

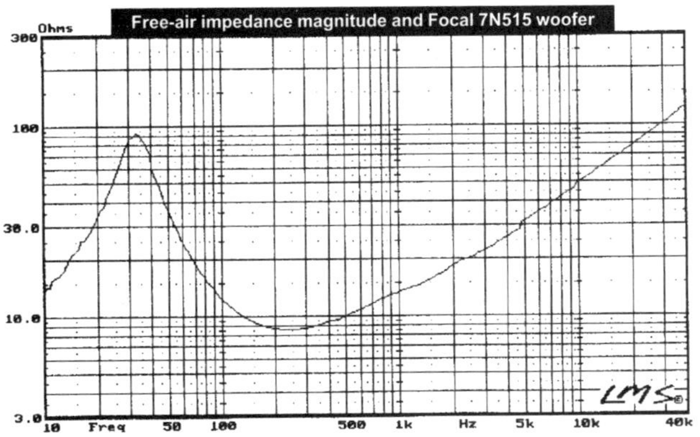

········ 7N515 E

FIGURE 3.1

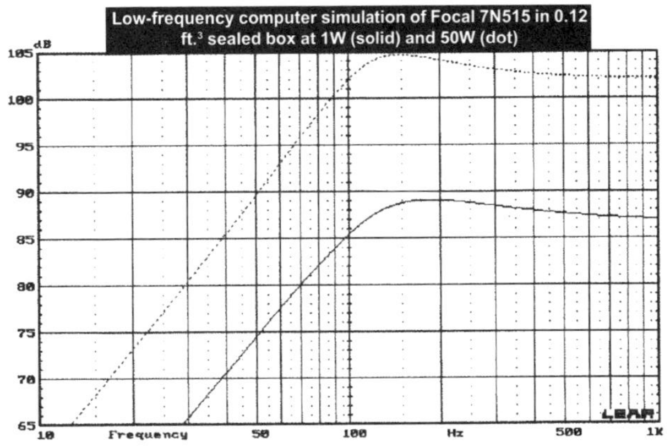

—— 7n515 .12cf 2.83v
········ 7n515 .12cf 50w

FIGURE 3.2

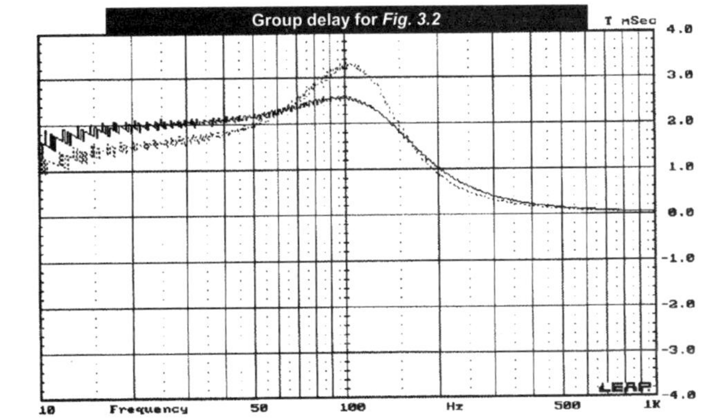

—— 7n515 .12cf 2.83v
········ 7n515 .12cf 50w

FIGURE 3.3

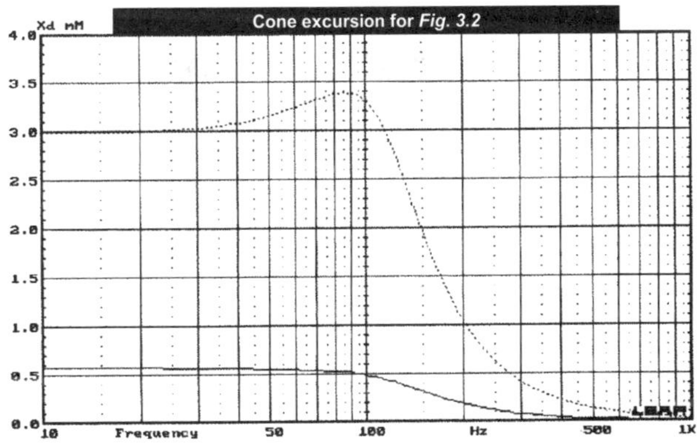

—— 7n515 .12cf 2.83v
········ 7n515 .12cf 50w

FIGURE 3.4

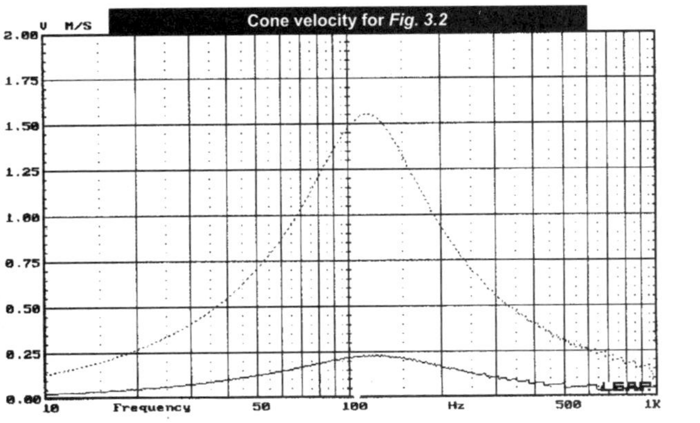

—— 7n515 .12cf 2.83v
········ 7n515 .12cf 50w

FIGURE 3.5

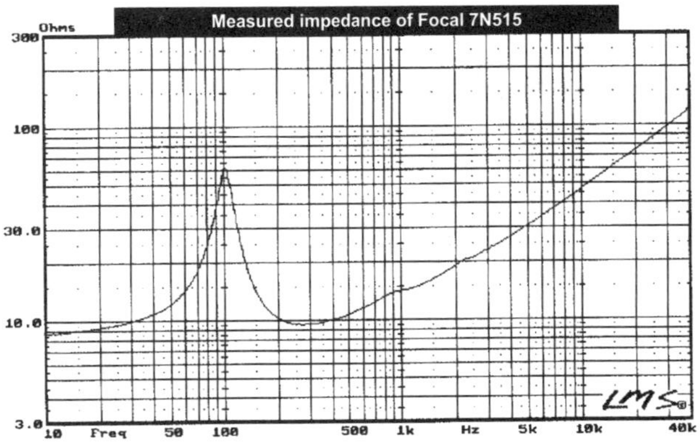

—— Mod3a

FIGURE 3.6

where it maintains its level until a dip in the response occurs at 2kHz, finally rolling off fairly steeply above 4kHz. The crossover must begin contouring the woofer's response at 250-300Hz, giving the final design an 86-87dB efficiency level. *Figure 3.10* shows the two 7N515 samples to have as well-matched frequency response curves as they have T/S parameters.

3.30 The Subwoofer

At first I planned to use a bandpass design with the 7N515/T120 combination. The prototype

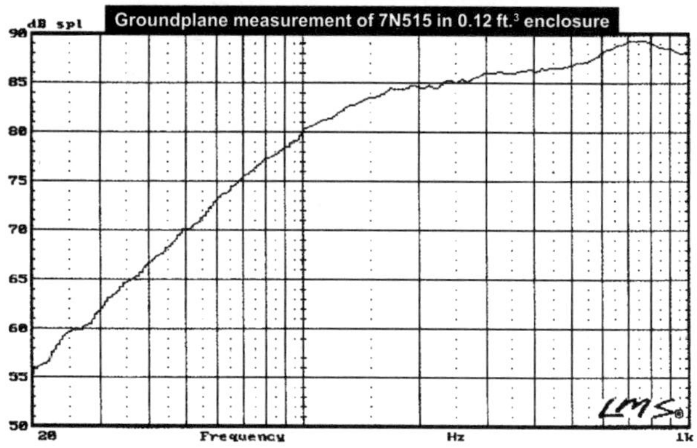

— Focal 7N515

FIGURE 3.7

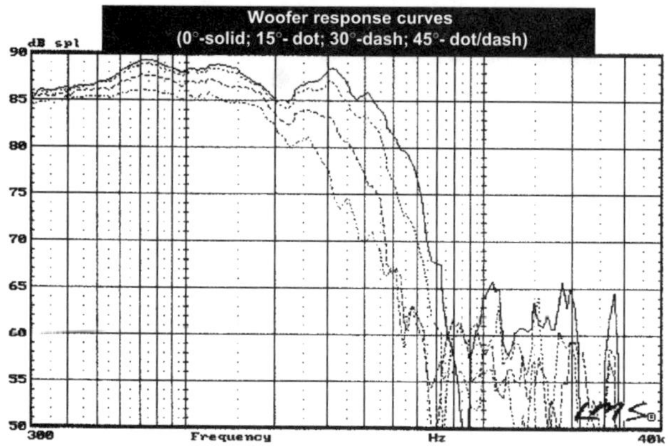

— Woof 0 a
······· Woof 15 a
- - - Woof 30 a
-·-· Woof 45 a

FIGURE 3.8

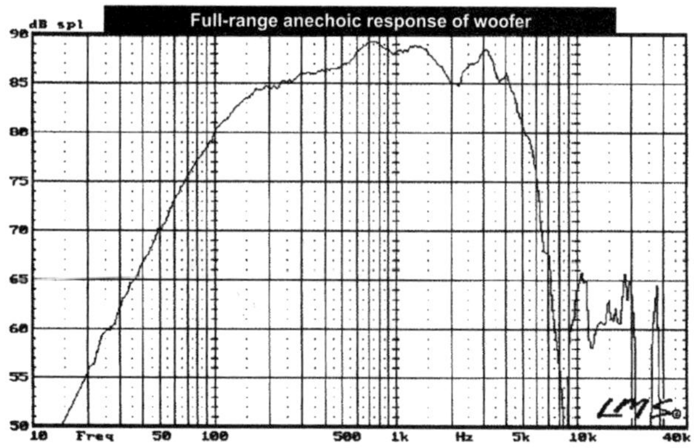

— Focal 7N515

FIGURE 3.9

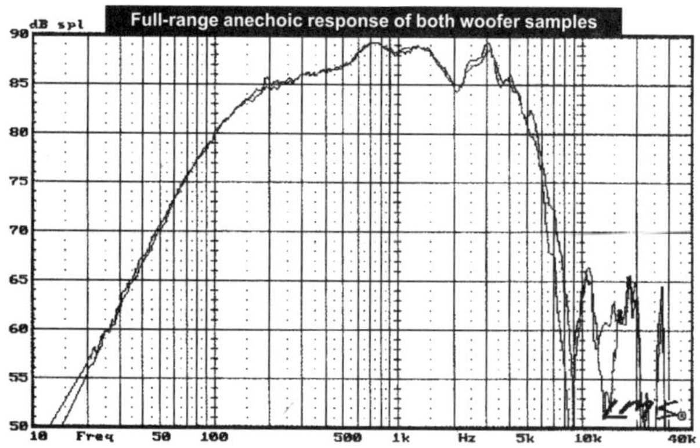

— Focal 7N515 B
······· Focal 7N515 A

FIGURE 3.10

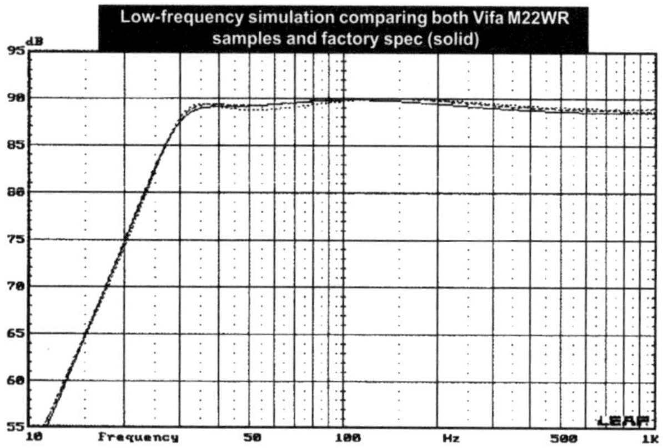

— Vifa factory
······· Vifa sample b
- - - Vifa 8" sub 1.4cf 1 watt

FIGURE 3.11

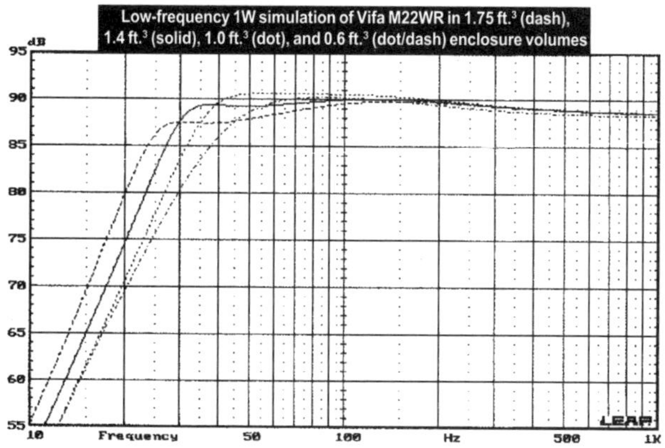

— Vifa 8" sub 1.4cf 1 watt
······· Vifa 8" sub 1cf 1 watt
- - - Vifa 8" sub 1.75cf 1 watt
-·-· Vifa 8" QB3 .6cf 1 watt

FIGURE 3.12

worked well and had a low f_3 of 20Hz, but the driver manufacturer cancelled production during the long course of writing this book. The cancelled prototype's only other drawback was its 2.5 ft.^{3} box size. Starting over, I hoped to find a driver with good low-end performance, and one which would work in a more compact enclosure, perhaps no larger than 1.5 ft.^{3}. Fortunately I received several product data sheets from Vifa, featuring a new 8" subwoofer driver.

Although I prefer low-Q sealed boxes for this type of design, I was impressed with Vifa's 8" M22WR capabilities (intended for vented box use) and its promised f_3 of 30Hz. The T/S parameter test results, published in the September, 1992 issue of *Voice Coil*, are as follows:

	A	**B**	**Factory**
Z_{NOM}	6	6	6
R_E (Ω)	4.2	4.24	4.3
S_d (M^2)	0.0206	0.0206	0.224
Bl (TM)	8.05	8.59	8.8
V_{AS} (ltr)	84.1	82.3	55
C_{MS} (µM/N)	1396	1365	...
f_S (Hz)	24.7	23.9	30
Q_{MS}	2.048	1.963	2.73
Q_{ES}	0.299	0.273	0.38
Q_{TS}	0.261	0.239	0.33
SPL (2.83V)	89.3dB	89.4dB	88
L_E (1kHz)	0.82mH	0.81mH	1.1
(20kHz)	0.32mH	0.32mH	...

While the differences between the two samples and between the samples and the factory specs may seem significant, the low-frequency simulation of all three sets in *Fig. 3.11* shows the net result to be nearly identical. As long as certain ratios such as f_S/Q_{TS} remain fairly constant, small shifts in the T/S numbers are not significant.

I did computer simulations to determine the most suitable box size and tuning. I used four different box volumes, two determined by LEAP's Quick Cabinet Utility (a 0.6 ft.^{3} QB3 and 1.4 ft.^{3} EBS) , and two from the Vifa literature (1 ft.^{3} and 1.75 ft.^{3}). The 1W simulations in *Figs. 3.12-3.18* illustrate the frequency response, group delay, cone excursion, cone velocity, port excursion, port velocity, and impedance respectively. Looking at the frequency response curves, the QB3 is obviously not acceptable since it has an f_3 of only 42Hz, which, although the sound would be very well-damped and the enclosure compact, I really wanted to achieve something closer to an f_3 of 30Hz. The f_3 for the 1 ft.^{3} enclosure is 35Hz and 28Hz for the 1.4 ft.^{3} box. Since the group delay curve for the 1.75 looked much too unattractive, the choice was between 1 ft.^{3} and 1.4 ft.^{3}.

To make a final decision I did another set of simulations at 38W input for 1.75 ft.^{3}, 1.4 ft.^{3} and 1 ft.^{3} box volumes shown in *Figs. 3.19-3.21* for the frequency response, group delay, and cone excursion. The output for all three is about 103dB, but the cone excursion for the 1.75 ft.^{3} enclosure exceeds the X_{MAX} + 15% (7.5mm for the 6.5mm X_{MAX} of the M22WR) criteria. The 1.4 ft.^{3} box is still linear at this power level down to about 25Hz, and since it provides a lower cutoff than the 1 ft.^{3}

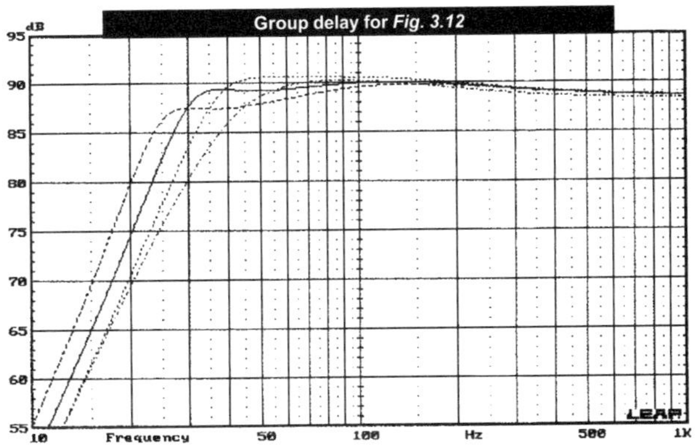

—— Vifa 8" sub 1.4cf 1 watt
······· Vifa 8" sub 1cf 1 watt
- - - Vifa 8" sub 1.75cf 1 watt
-·-· Vifa 8" QB3 .6cf 1 watt

FIGURE 3.13

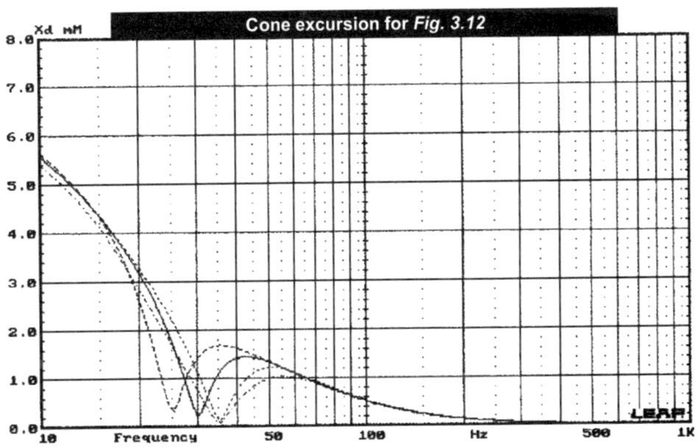

—— Vifa 8" sub 1.4cf 1 watt
······· Vifa 8" sub 1cf 1 watt
- - - Vifa 8" sub 1.75cf 1 watt
-·-· Vifa 8" QB3 .6cf 1 watt

FIGURE 3.14

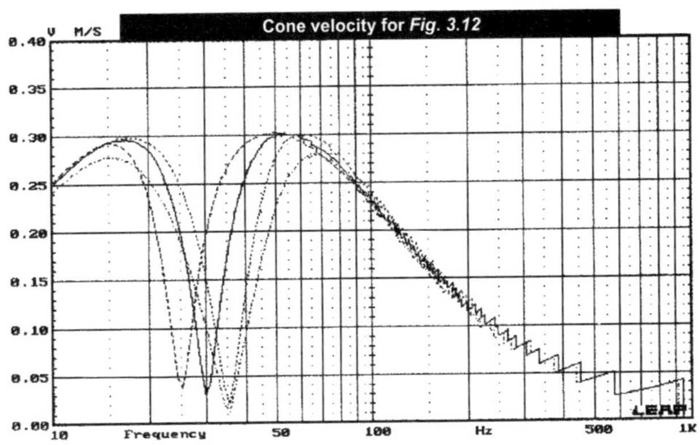

—— Vifa 8" sub 1.4cf 1 watt
······· Vifa 8" sub 1cf 1 watt
- - - Vifa 8" sub 1.75cf 1 watt
-·-· Vifa 8" QB3 .6cf 1 watt

FIGURE 3.15

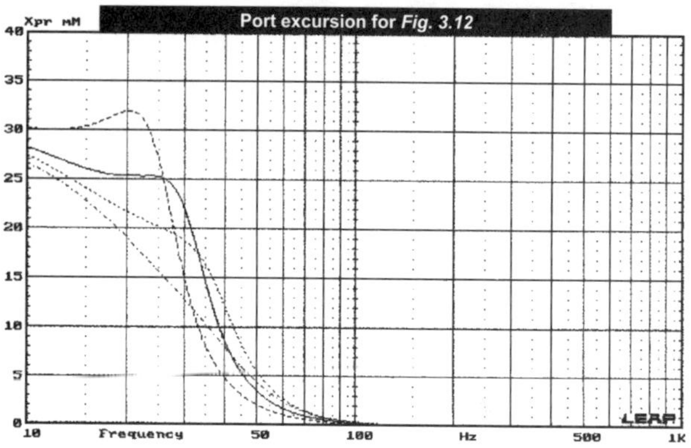

— Vifa 8" sub 1.4cf 1 watt
········ Vifa 8" sub 1cf 1 watt
- - - Vifa 8" sub 1.75cf 1 watt
-·-· Vifa 8" QB3 .6cf 1 watt

FIGURE 3.16

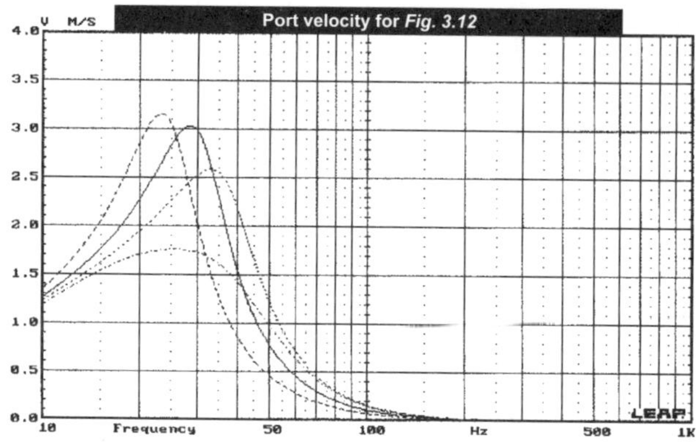

— Vifa 8" sub 1.4cf 1 watt
········ Vifa 8" sub 1cf 1 watt
- - - Vifa 8" sub 1.75cf 1 watt
-·-· Vifa 8" QB3 .6cf 1 watt

FIGURE 3.17

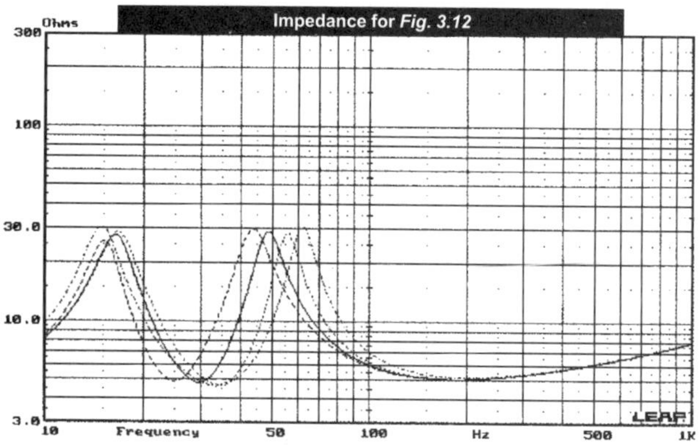

— Vifa 8" sub 1.4cf 1 watt
········ Vifa 8" sub 1cf 1 watt
- - - Vifa 8" sub 1.75cf 1 watt
-·-· Vifa 8" QB3 .6cf 1 watt

FIGURE 3.18

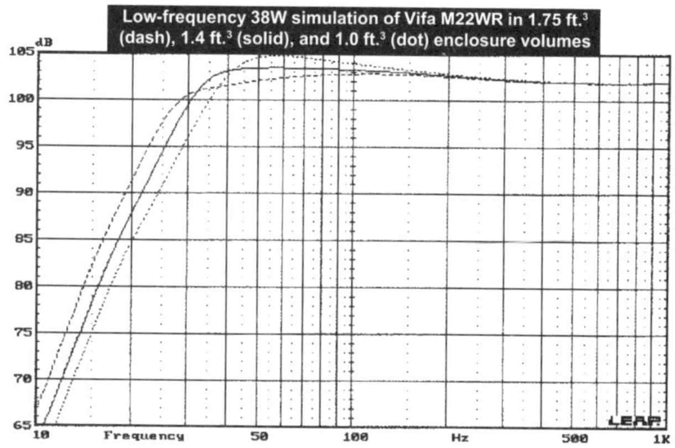

— Vifa 8" sub 1.4cf 38w
········ Vifa 8" sub 1cf 38w
- - - Vifa 8" sub 1.75cf 38w

FIGURE 3.19

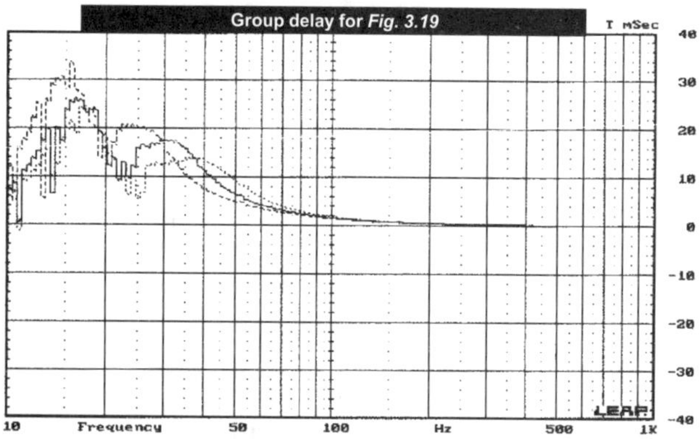

— Vifa 8" sub 1.4cf 38w
········ Vifa 8" sub 1cf 38w
- - - Vifa 8" sub 1.75cf 38w

FIGURE 3.20

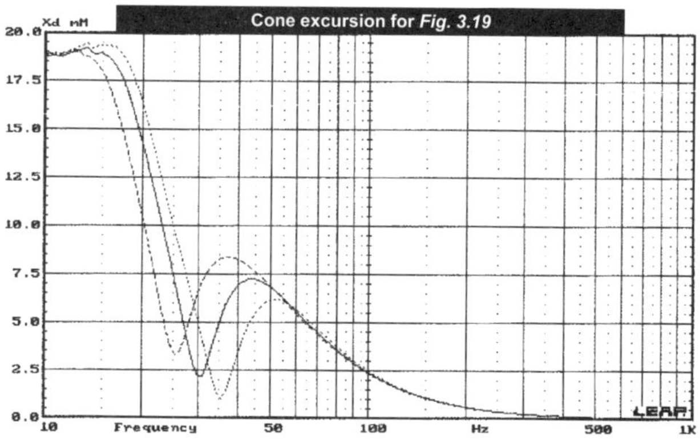

— Vifa 8" sub 1.4cf 38w
········ Vifa 8" sub 1cf 38w
- - - Vifa 8" sub 1.75cf 38w

FIGURE 3.21

box, I chose it. Vent diameter for all the simulations is 3", since it is the largest diameter which will tune the enclosure volume and be short enough to fit into a reasonable box dimension.

Once our prototype box was ready, I verified the woofer low-frequency performance by doing an outdoor groundplane measurement illustrated in *Fig. 3.22*. The measured f_3 was 31Hz, without any damping material in place. Final measurements with damping material were about $f_3 = 29Hz$. This is acceptably close to the simulations. The high frequency measurements are shown in *Fig. 3.23.* Although I had no intention of using this woofer above 120Hz, the graph shows it would be difficult to use it with a crossover frequency above 1.5kHz.

The woofer was mounted in the enclosure so it would be about 11" above the floor. By mounting the terminals on the bottom of the box, the user is left with a variety of options. Coupling to the room can be adjusted by placing the box out into the room away from the walls or sideways loading along the wall. It is even possible to experiment with firing the woofer straight up.

It should also be noted that the term subwoofer as applied here is actually plural: subwoofers. Although it's feasible to use a summing electronic crossover (available on both the TDM and Marchand units), my choice was a subwoofer for each channel, a superior setup in my opinion. I placed each subwoofer outside each satellite; that is, the left sub is to the left of the left satellite, with the right channel arranged in the same fashion. The satellites are placed 6'-8' apart.

Peter W. Mitchell discussed the superiority of two subs over a single unit in *Stereophile.*[1] Mitchell believes the increased radiating area plus the new digital recording practices, make better sound possible with two subs rather than a single, summed channel. Before the advent of digital recording, many engineers summed the frequencies below 50Hz to conserve space on long-play discs. Compact disc technology allows unlimited recording in the bass frequencies, with cutoff points set at the low end of human hearing, about 20Hz.

The general assumption has always been that any signal below 100Hz is interpreted by human ears as nondirectional and that the brain assigns its location to the nearest set of upper harmonics (your satellite speaker). While this psychoacoustic phenomenon certainly exists, speaker crossovers do not have infinite attenuation at the crossover frequency. Harmonic information is still present to some degree even with steep slopes set to the 80-120Hz. range, and can affect the stereo image. For these reasons I strongly recommend two subwoofers if your listening room has enough space.

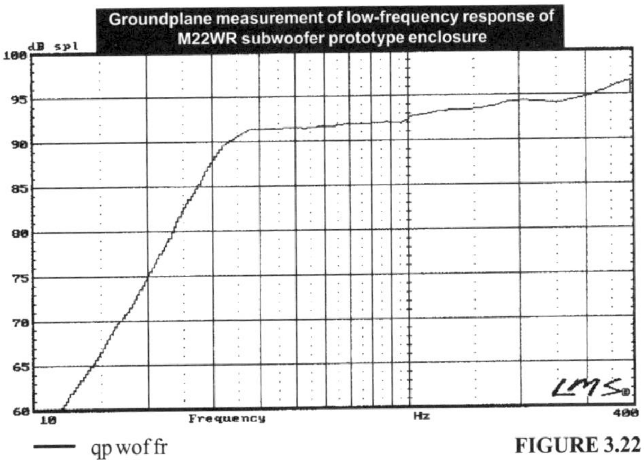

FIGURE 3.22

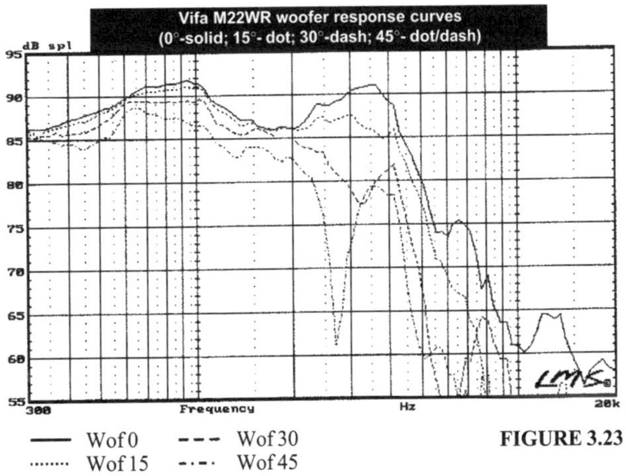

FIGURE 3.23

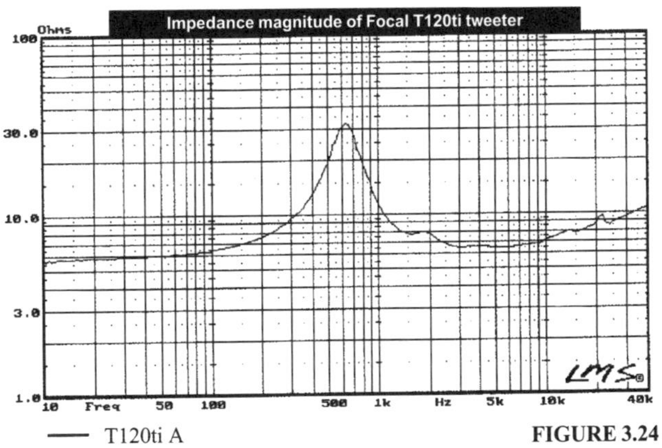

FIGURE 3.24

3.40 The Tweeter

The Focal T120ti is the latest version of the popular T120 tweeter. Although earlier Kevlar versions of this 1" inverted dome tweeter encountered a little criticism by some of the audiophile press (which I believe was unjustified), the new titanium versions have been well received. Only slightly more expensive than the Scan-Speak D2905, these tweeters are excellent performers.

The T120ti is a "big" tweeter with a larger than usual faceplate (about 4.75" square) and a 25 oz. magnet. This magnet size is relatively large since most tweeters have 8-10 oz. magnets and 25 oz. magnets are generally found on 8-10" woofers.

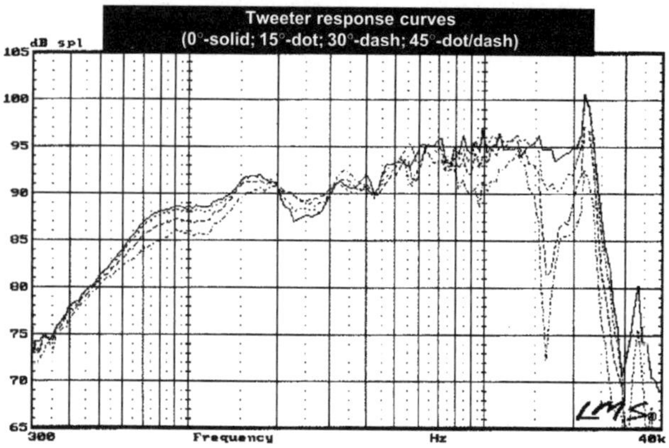

— Focal T120ti
······· Tw 15
- - - Tw 30
-·-· Tw 45

FIGURE 3.25

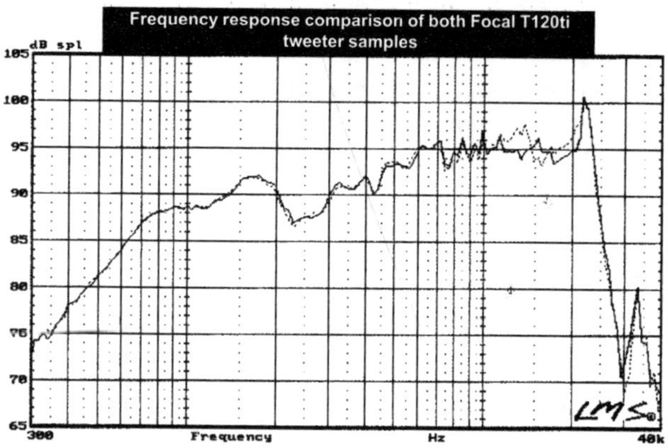

— Focal T120ti A
······· Focal T120ti B

FIGURE 3.26

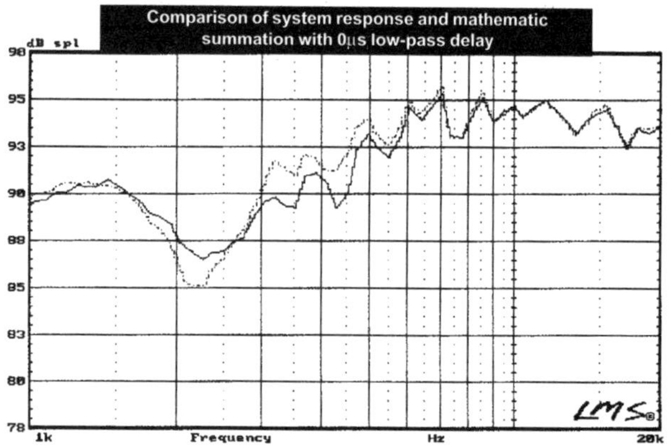

— Sys 0 net
-·-· Sum 0 delay

FIGURE 3.27

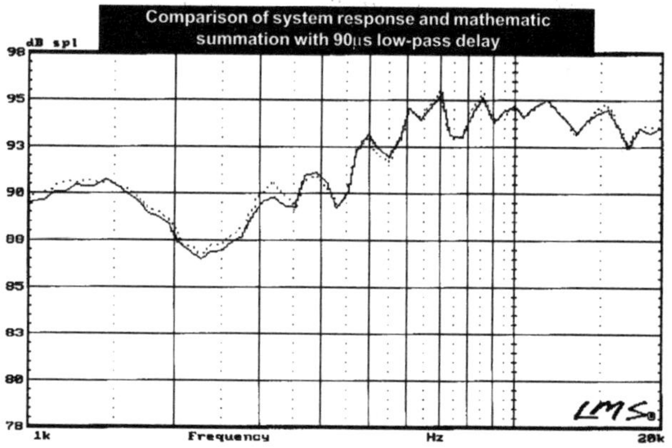

— Sys 0 net
- - - Sum 90μs delay

FIGURE 3.28

This tweeter's impedance curve (*Fig. 3.24*) is obviously undamped by magnetic fluid, since the resonance rises to about 32Ω. I used this curve to calculate the following parameters:

Z_{NOM}	8Ω
R_E	6.07Ω
f_S	640Hz
Q_{MS}	2.78
Q_{ES}	0.64
Q_{TS}	0.52
L_E (1kHz)	0.091mH
(20kHz)	0.023mH

The response at 0°, 15°, 30°, and 45° is depicted in *Fig. 3.25*, where the tweeter exhibits a rising response from 2.5kHz to about 6kHz and then remains flat ±1.5dB to 20kHz. The off-axis response characteristics of the inverted dome is very good out to 30°, where it remains flat to 15kHz. Both samples, shown in *Fig. 3.26*, were well matched to ±0.25dB out to 12kHz, with 1-2dB of variation from 12-25kHz. Focal recommends this driver be crossed at a minimum of 3kHz, which is primarily due to its response characteristics. Contouring this tweeter for a low 2kHz crossover without losing significant efficiency would be difficult.

3.50 The Crossover Configurations

The physical time delay between the Focal 7N515 and the T120ti mounted flush on the same baffle measures about 1.3", or about 95μs. This was confirmed using the same procedure described in *Chapter 1* and is illustrated in *Figs. 3.27-3.30* for a 0μs, 90μs, 95μs, and 100μs low-pass delay respectively.

Figure 3.31 shows the simulated summation of two acoustically-flat fourth-order L-R high-pass and low-pass sections with the low-pass section

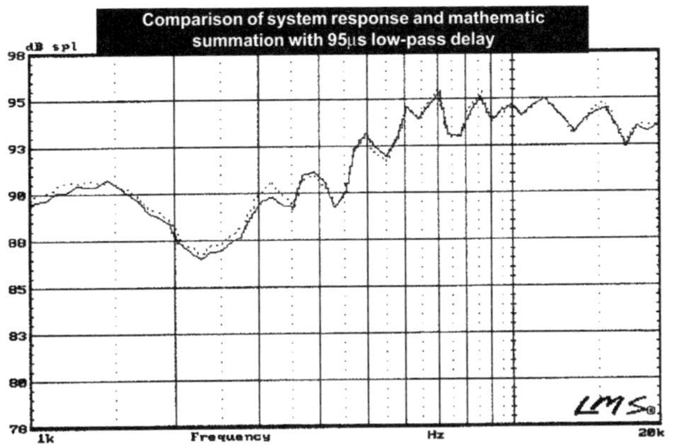

— Sys 0 net
-·-· Sum 95us delay

FIGURE 3.29

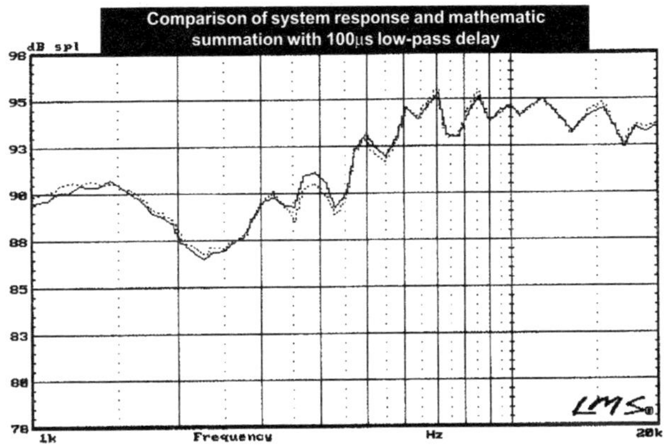

— Sys 0 net
······ Sum 100us delay

FIGURE 3.30

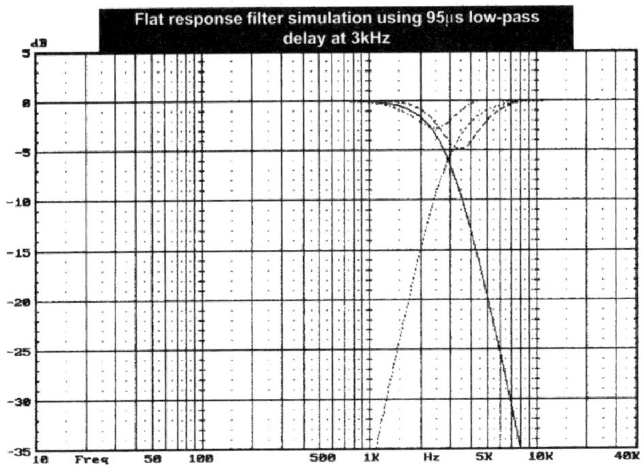

— Lo Pass Section Only: Magnitude
······ Hi Pass Section Only: Magnitude
- - - Sum= Lo+Hi (InPhase): Magnitude
-·-· Sum= Lo-Hi (Reverse): Magnitude

Lo Filtr: 4th-LnkRly Hi Filtr: 4th-LnkRly
Freq = 3000.00 Hz Freq= 3000.00 Hz
Level= 0.00 dB Level= 0.00 dB
Delay= 95.00µs Delay= 0.00 µs

FIGURE 3.31

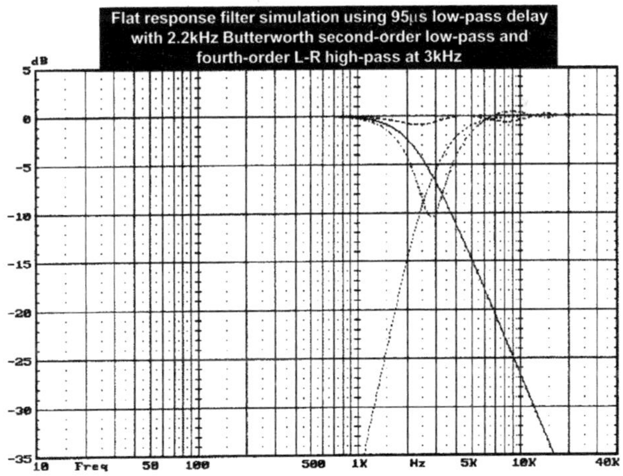

— Lo Pass Section Only: Magnitude
······ Hi Pass Section Only: Magnitude
- - - Sum= Lo+Hi (InPhase): Magnitude
-·-· Sum= Lo-Hi (Reverse): Magnitude

Lo Filtr: 4th-LnkRly Hi Filtr: 4th-LnkRly
Freq = 2200.00 Hz Freq= 3000.00 Hz
Level= 0.00 dB Level= 0.00 dB
Delay= 95.00µs Delay= 0.00 µs

FIGURE 3.32

delayed by 95µs at the minimum 3kHz crossover frequency of the tweeter. Both drivers for this configuration are 90-100° out-of-phase for either an in-phase or out-of-phase electrical connection. Unfortunately, the Focal 7N515 is not a good candidate for a 3kHz crossover. However a small amount of manipulation produced an asymmetrical crossover with a second-order Butterworth low-pass at 2.2kHz and a fourth-order L-R high-pass at 3kHz. The result easily provides a flat response with only a slight 1dB ripple (*Fig. 3.32*). This technique of using asymmetrical high-pass and low-pass sections to compensate for interdriver time delay has been suggested both by B. Hillerich[2] and M.O.J. Hawkesford[3], although the latter seemed to believe, unjustifiably I believe, that the technique was not applicable to high crossover frequencies. I will discuss asymmetrical crossovers later.

3.60 The Woofer Crossover

The full-range response of the 7N515 (*Fig. 3.9*) exhibits the same rising response typical of most woofers in small enclosures. The crossover must start contouring the woofer at about 300Hz and then produce a second-order Butterworth slope at 2.2kHz. The only "kink" in this design is the peak at 3kHz. When you use LC filters to contour and roll off a woofer, most of what you are doing is tilting the response and then shaping it at the

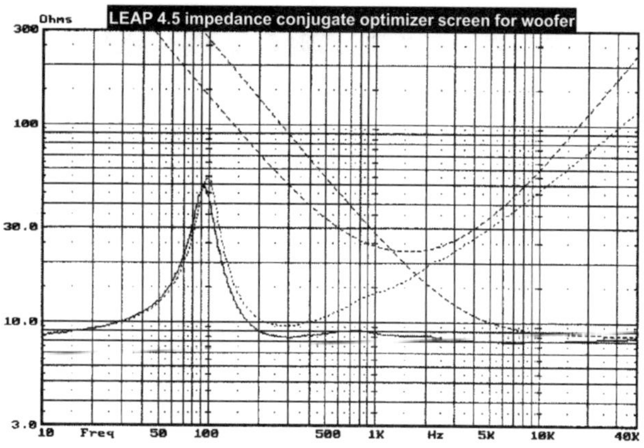

Combined Impedance
Original Impedance
Conjugate Nets
DGL Data Info
DGL# 2:Focal 7N515

FIGURE 3.33

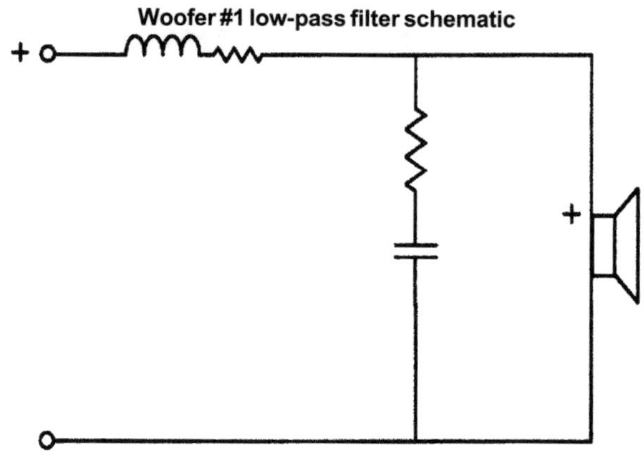

FIGURE 3.34

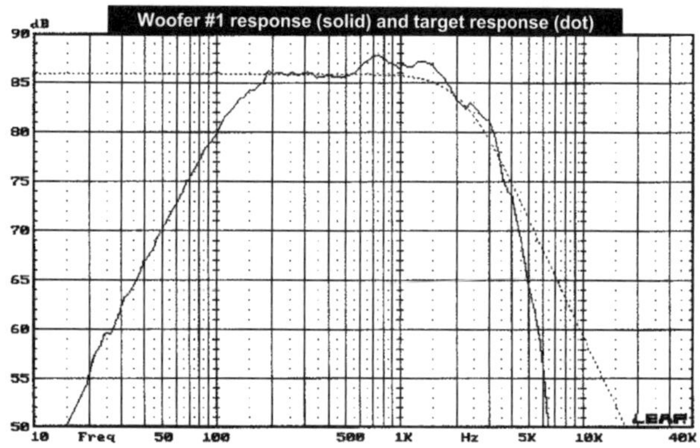

Wof sec 2 p12
Generic 1p

FIGURE 3.35

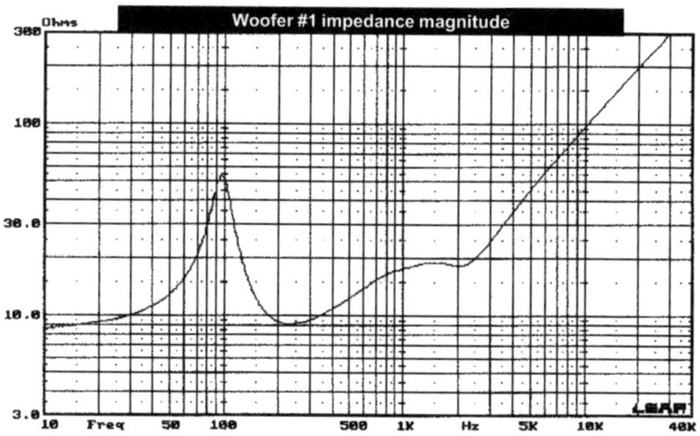

Wof sec 2 p12

FIGURE 3.36

upper end. This does not, however, compensate for any major anomalies you encounter along the way. If the 7N515 were flat in the region between 1.5kHz and 3kHz and did not have the dip at 2kHz, then the 3kHz "peak" would not require any special manipulation.

I followed the same procedure as before, and used the LEAP Conjugate Designer to calculate component values which would render the impedance flat to a nominal value. *Figure 3.33* depicts the calculated values:

LCR	CR
0.88mH	5.85μF
11.45μF	8.55Ω
22.84Ω	

I kept these values for reference, and devised three network topographies. The three network sections are:

1. First-order low-pass + CR conjugate.
2. Second-order low-pass + half-order low-pass + CR conjugate.
3. Half-order low-pass + LCR + first-order low-pass + CR conjugate.

I combined computer cut and try procedures and optimizations to find component values for each of these filter sections. This method initially turns loose all the circuit values by marking them "on," thus allowing the computer program to alter each value in its attempt to achieve the target response. Often one or more of the components will "explode" or rapidly increase to values which go far beyond practical sizes. You can usually manually "play" with these "sensitive" components until you find likely nominal values. They are then "fixed" or turned off. You can use a manual trial and error process to see which response outcome results from changing a

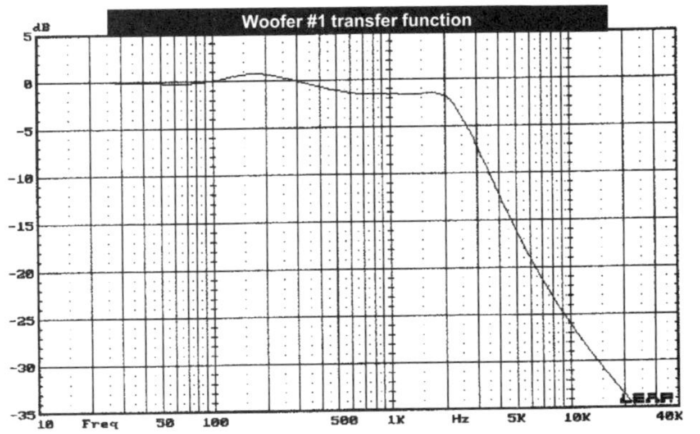

— Wof sec 2 p12

FIGURE 3.37

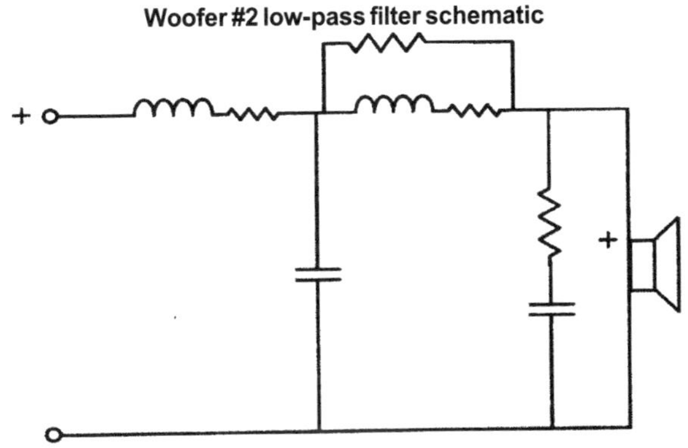

FIGURE 3.38

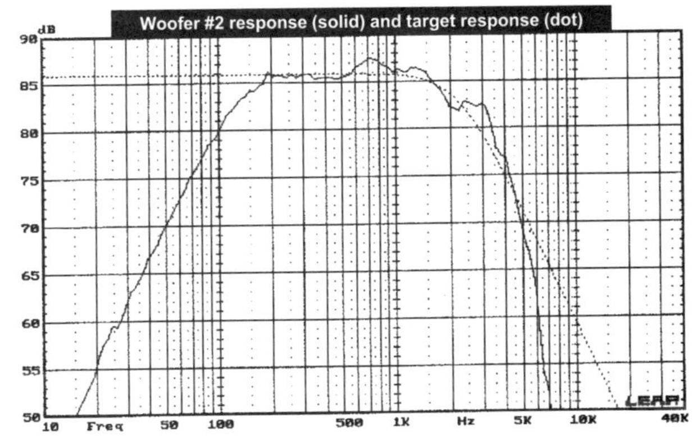

— Wof 2nd order 2500 p14
······· Generic lp

FIGURE 3.39

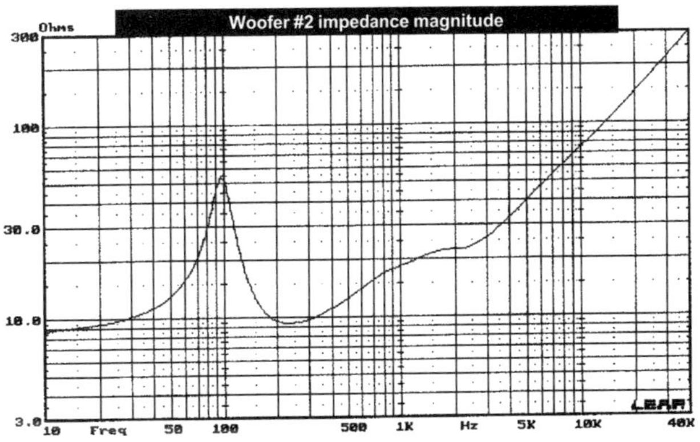

— Wof 2nd order 2500 p14

FIGURE 3.40

particular component value. When you get close to the target, you can usually turn on the optimizer for rapid results. The process varies for each network topography combination and driver, so no one method or answer will work in all cases. Sometimes just turning on all the values for optimization and then running the optimizer will rapidly achieve the final target.

My results and the circuit diagrams for each low-pass filter are shown in the following:

Woofer#	Circuit	Response +Target	Z	Transfer Function
1	3.34	3.35	3.36	3.37
2	3.38	3.39	3.40	3.41
3	3.42	3.43	3.44	3.45

All three circuits are compared in *Figs. 3.46-3.48.* Circuits #1 and #2 in *Fig. 3.46* could not be optimized to provide a response that tracked the target lower than -10dB. I selected woofer circuit #3, since it follows the target response to about -17dB, including the 3kHz peak filter. The woofer response, with and without the final network, is compared on-axis in *Fig. 3.49* and 30° off-axis in *Fig. 3.50.*

3.70 The Tweeter Crossover

The Focal T120ti on-axis response in *Fig. 3.26* shows the SPL between 6kHz and 20kHz to be about 95dB, meaning that the tweeter must be attenuated by 9dB to match the 86dB woofer level. The rising response between 3kHz and 6kHz closely matches the "knee" of a fourth-order L-R filter at 3kHz, so requires no additional contouring. Although the tweeter resonance is substantial, 35Ω at 632Hz, I did not add conjugate circuits to compensate for the resonance in the attempted networks, partially because the L-pads I used tend to "swamp" the resonance. The analysis (*Fig. 3.51*) produced these values for a resonant LCR:

L = 0.76mH
C = 83.6μF
R = 7.7Ω

Using the 86dB SPL target, I did optimizing routines for four high-pass circuits to a fourth-

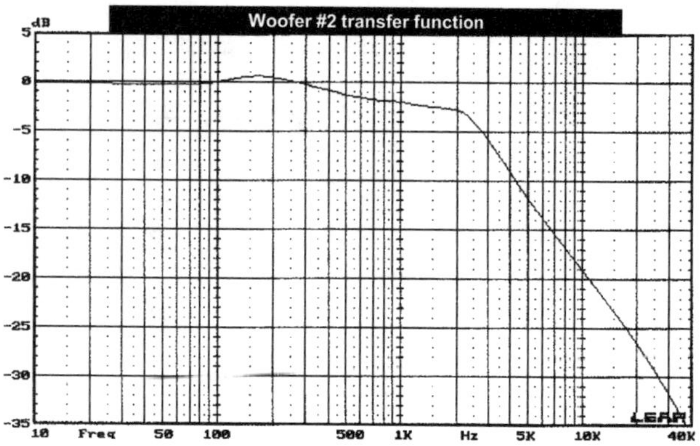

— Wof 2nd order 2500 p14

FIGURE 3.41

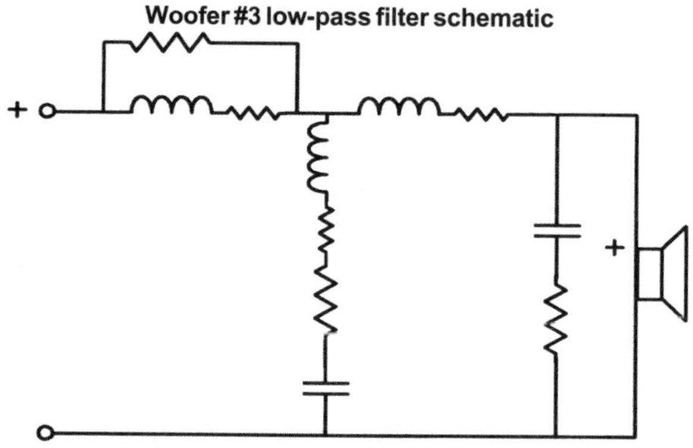

FIGURE 3.42

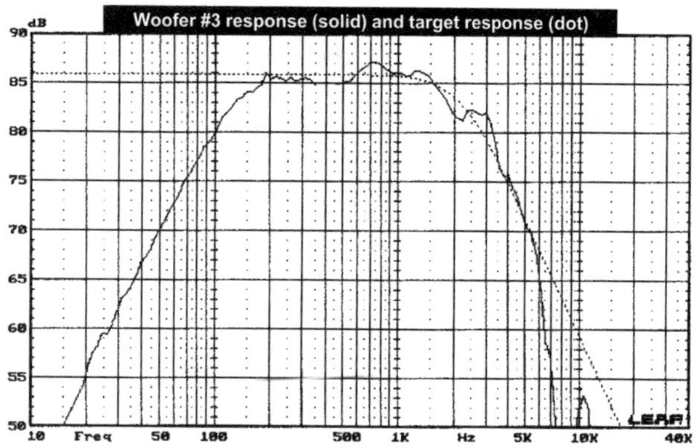

— Wof 2nd at 2200 p20
.... Generic 1p

FIGURE 3.43

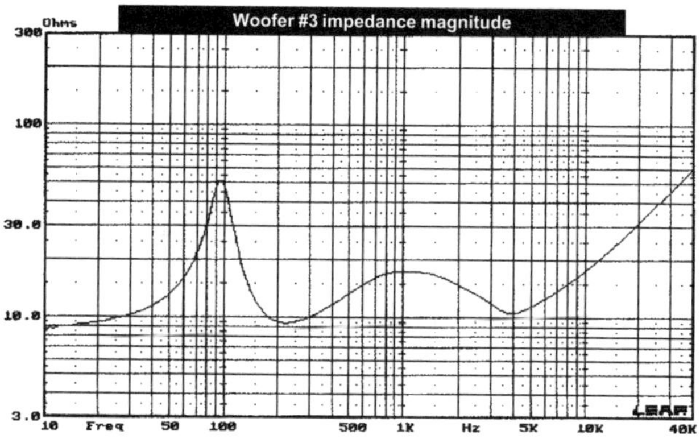

— Wof 2nd at 2200 p20

FIGURE 3.44

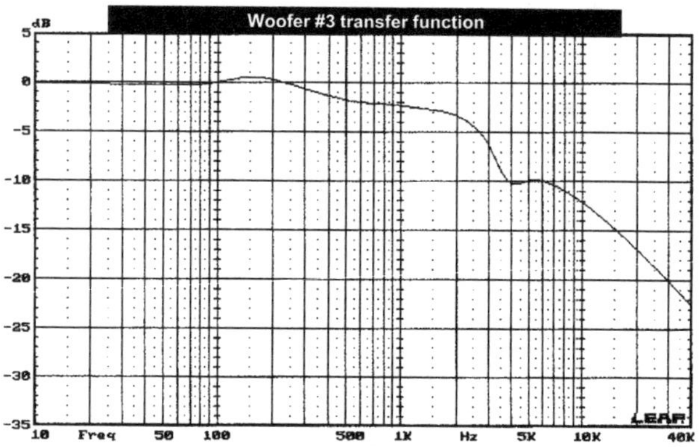

— Wof 2nd at 2200 p20

FIGURE 3.45

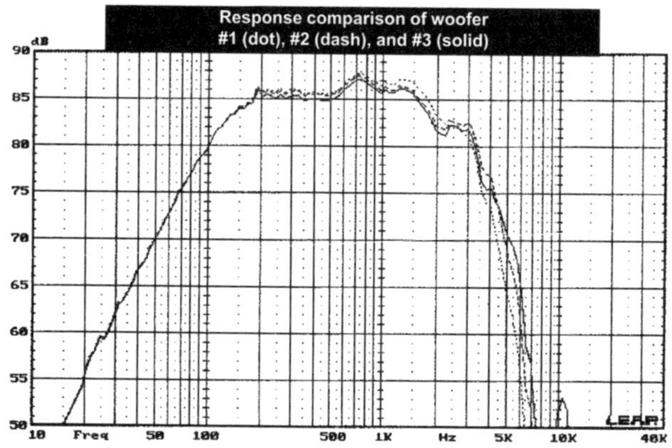

— Wof 2nd at 2200 p20 b
.... Wof sec 2 p12
- - - Wof 2nd order 2500 p14

FIGURE 3.46

order L-R high-pass response at 3kHz. These network sections are as follows:

1. Second-order high-pass + L-pad attenuation.
2. Third-order high-pass + series attenuation resistor.
3. Fourth-order high-pass + L-pad attenuation.
4. Third-order high-pass + contour + series attenuation resistor.

The four high-pass filters are depicted in the following diagrams:

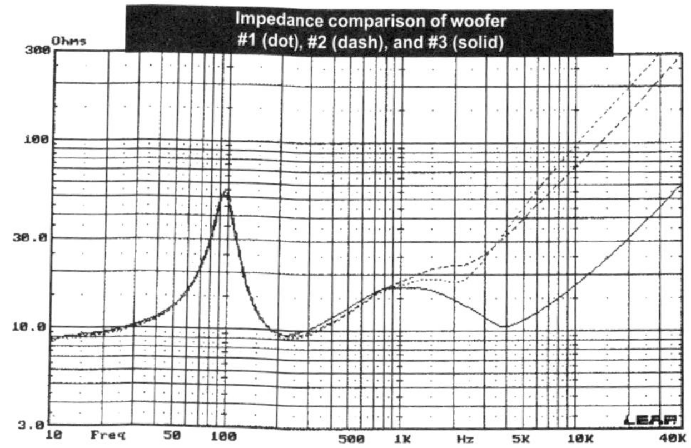

— Wof 2nd at 2200 p20 b
······· Wof sec 2 p12
- - - Wof 2nd order 2500 p14

FIGURE 3.47

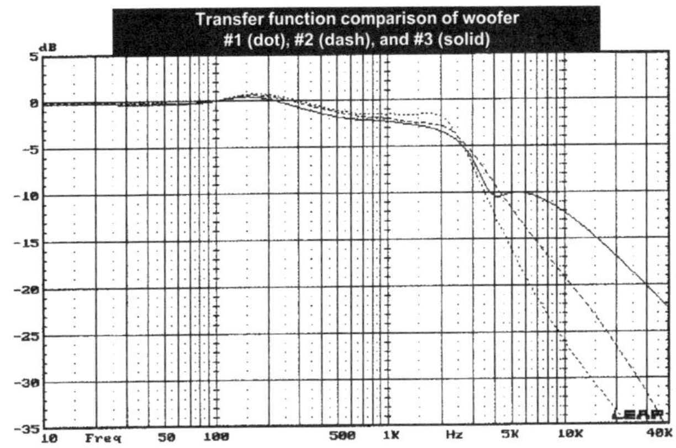

— Wof 2nd at 2200 p20 b
······· Wof sec 2 p12
- - - Wof 2nd order 2500 p14

FIGURE 3.48

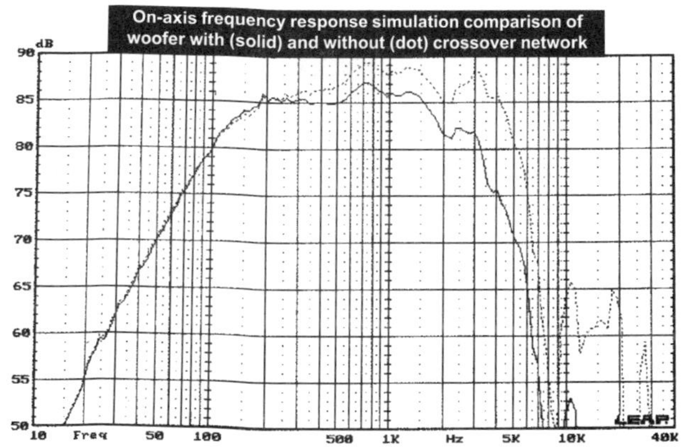

— Wof 2nd at 2200 p20 b
······· Focal 7N515

FIGURE 3.49

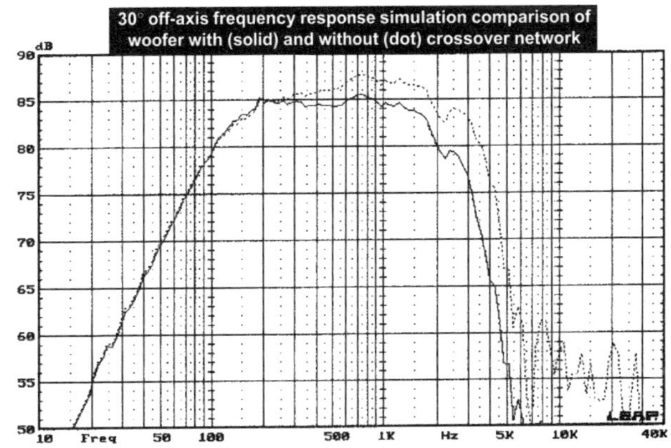

— Wof 2nd at 2200 p20 b
······· Focal 7n515

FIGURE 3.50

Tweeter#	Circuit	Response +Target	Z	Transfer Function
1	3.52	3.53	3.54	3.55
2	3.56	3.57	3.58	3.59
3	3.60	3.61	3.62	3.63
4	3.64	3.65	3.66	3.67

All four networks are compared simultaneously in *Figs. 3.68-3.70*. Each of these high-pass realizations would probably have been usable, since the transfer functions were all similar and none of the circuits caused excessively low impedance minima. Circuit #4 used a series half-order filter to contour the tweeter, which I tend to avoid in tweeter design. Circuit #3 called for a large capacitor value in the CR contour circuit (8μF), and the second inductor in circuit #2 was over 2mH. Comparing the 10dB scale response plot in *Fig. 3.71*, circuit #1 clearly provides the least attenuation in the second octave below the crossover frequency.

3.80 System Analysis

I chose the tweeter circuit by comparing the in- and out-of-phase summations for each of the four tweeter networks in woofer network #3:

Tweeter#	In-phase	Out-of-phase
1	3.72	3.73
2	3.74	3.75
3	3.76	3.77
4	3.78	3.79

All four circuit combinations look usable (*Fig. 3.80*), except the woofer #3 and tweeter #1 combination provide the most nearly flat system response and the best null profile. I chose to use tweeter circuit #1, mostly because it was the simplest of the group with the best response. *Figures 3.81* and *3.82* compare the tweeter with and without the network on- and 30° off-axis.

The final summed simulations of woofer circuit #3 and tweeter #1 are displayed in *Figs. 3.83-3.86*

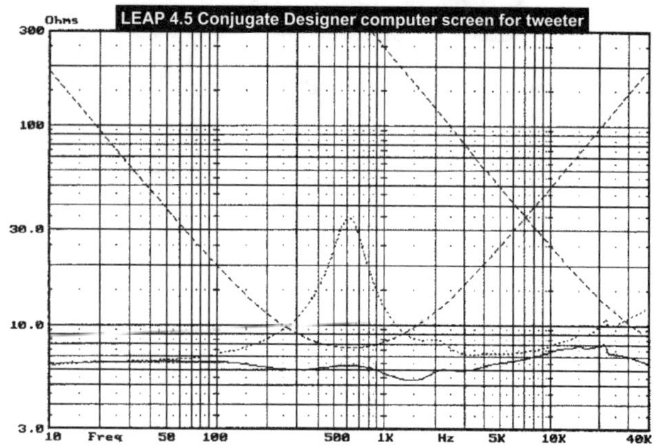

— Combined Impedance
······· Original Impedance
- - - Conjugate Nets

DGL Data Info
DGL# 2:Focal T120ti

FIGURE 3.51

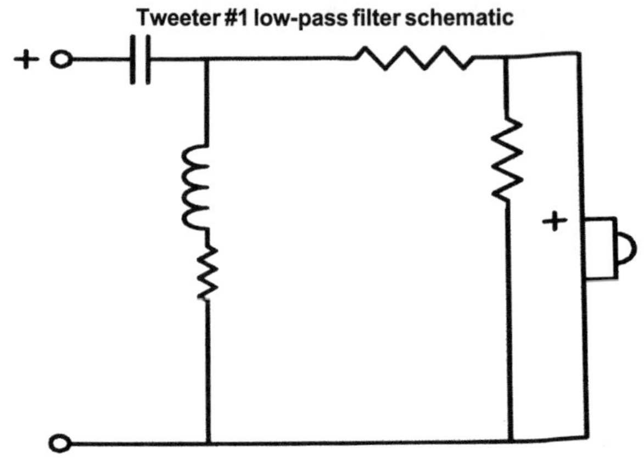

FIGURE 3.52

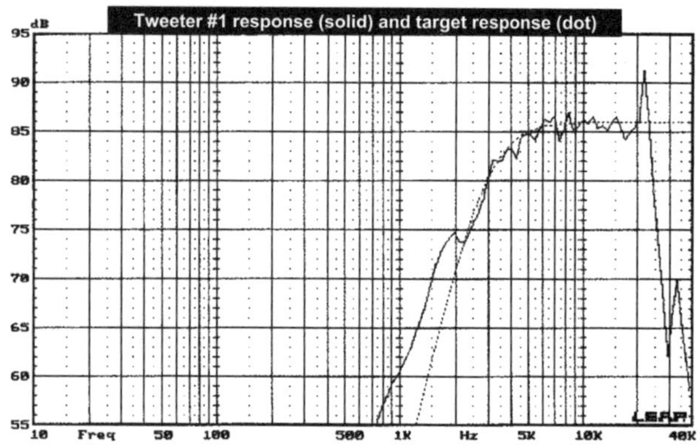

— Tw sect 4 + lpad p7
······· Generic Hp

FIGURE 3.53

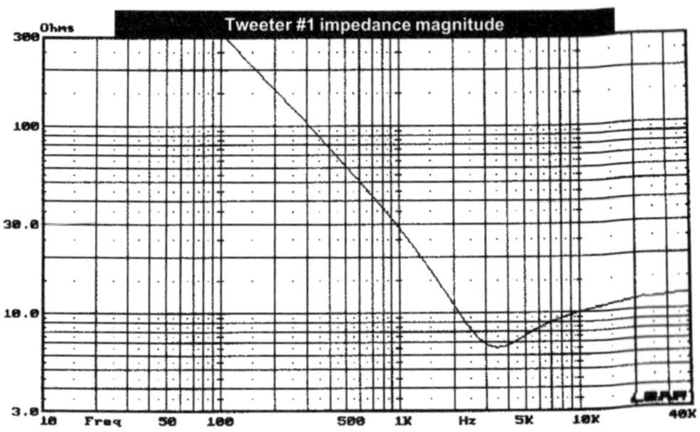

— Tw sect 4 + lpad p7

FIGURE 3.54

for the on-axis, 30° off-axis, impedance, and the combination's transfer functions. The final crossover schematic is in *Fig. 3.87*, with component values in *Table 3.1.*

TABLE 3.1
Crossover Component Values

Low-pass section

Value	Source
C1 6.47µF SCR poly (6µF + 0.47µF)	*
C2 7µF SCR poly (6µF + 1µF)	*
L1 1.2mH, 0.25Ω air core (12ga)	-
L2 0.4mH, 0.12Ω air core (14ga)	-
L3 0.56mH, 0.15Ω air core (12ga)	-
R1 6.8Ω, 15-25W	
R2 8Ω, 10W	
R3 5.1Ω, 10W	

High-pass section

C3 5µF SCR poly (4.7µF + 0.33µF)	*
L4 0.5mH, 0.13Ω air core (14ga)	-
R4 8Ω 6W (two 3W 16Ω)	~
R5 7Ω 6W (two 3W 14Ω)	~

*Orca Design & Manufacturing

-North Creek Music Systems

~Metal film resistors

3.90 Measured Performance

Figure 3.88 illustrates the final full-range on-axis frequency response of the Focal 7N515/T120ti prototype speaker. The response is ±2dB from 147Hz-20kHz, and ±3dB from 115Hz-22kHz, fitting well into the design criteria initially established. This is also close to the final computer

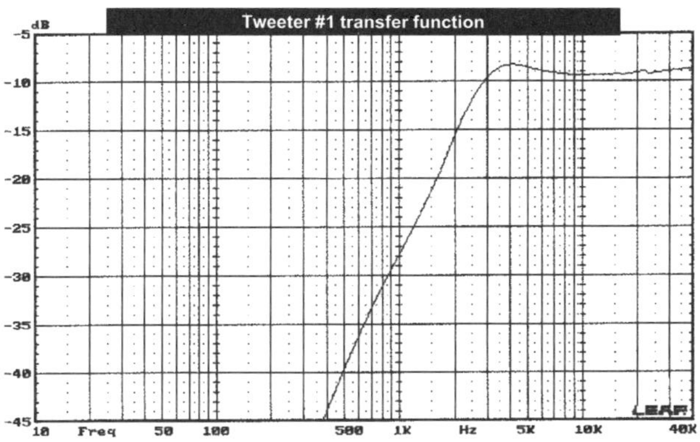

— Tw sect 4 + lpad sp7

FIGURE 3.55

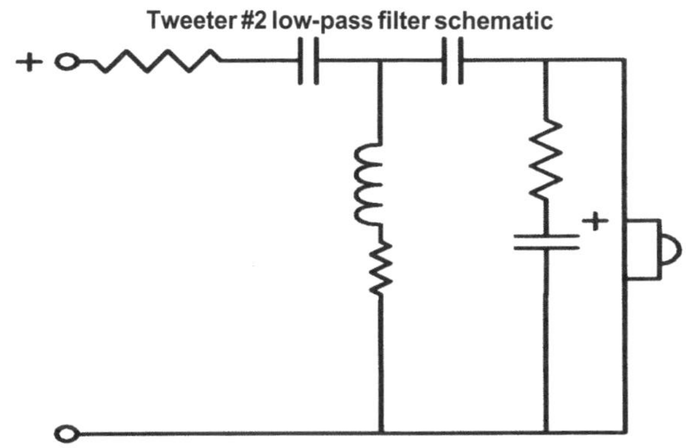

FIGURE 3.56

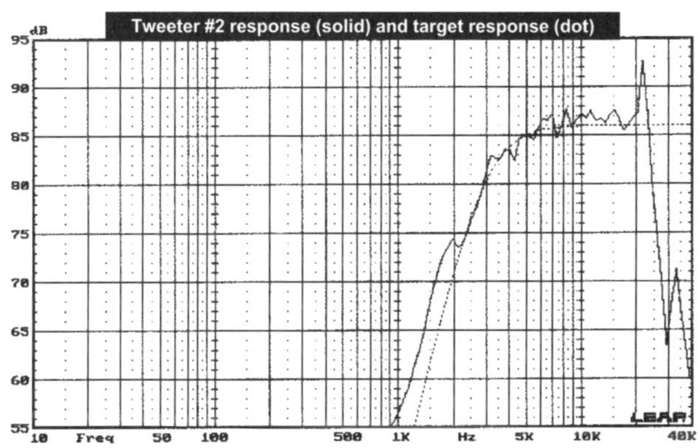

— Tw sect 1 + lpad p5
.... Generic Hp

FIGURE 3.57

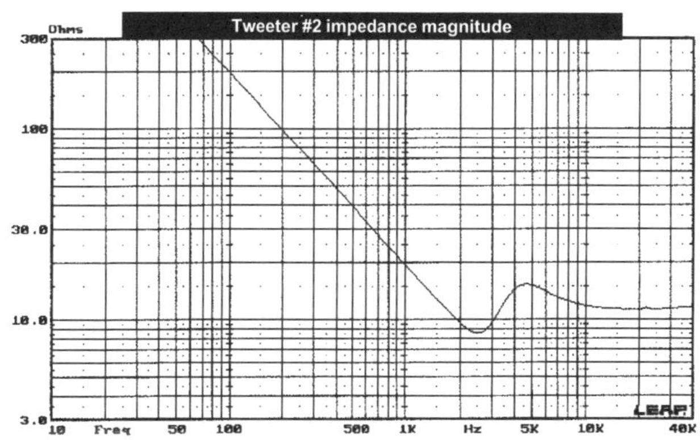

— Tw sect 1 + lpad p5

FIGURE 3.58

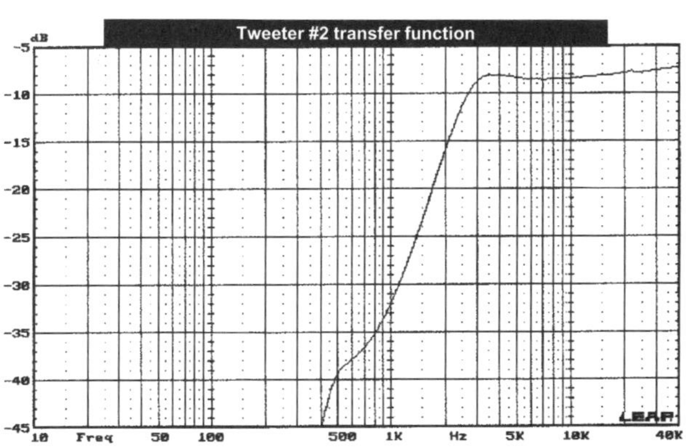

— Tw sect 1 + lpad p5

FIGURE 3.59

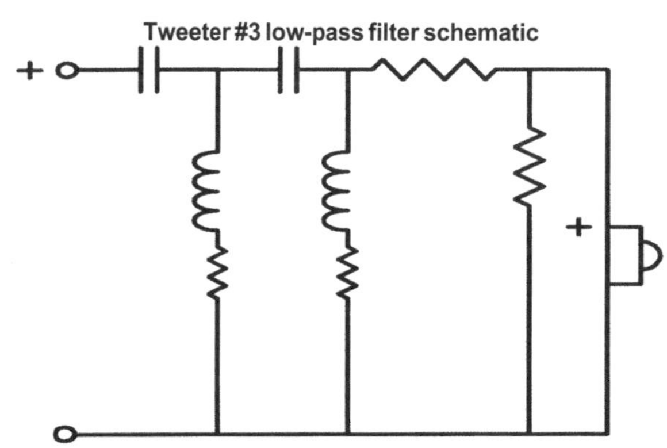

FIGURE 3.60

simulation in *Fig. 3.83*. The off-axis response at 0°, 15°, and 30° depicted in *Fig. 3.89* shows the speaker to stay in the ±2dB window out to 15kHz even at 30° off-axis. This profile degenerates slightly out to 45°, seen in *Fig. 3.90*, but the response is still quite flat. Even looking at the performance out to 90°, shown in *Fig. 3.91*, the response profile maintains its uniformity remarkably well with this design.

The vertical off-axis response reveals what I referred to earlier as the "ramifications" of using an asymmetrical crossover network. The vertical response curves taken up and over the top of the speaker (*Fig. 3.92*) show a deeper null at 15° than at 30°, while the opposite occurs for the down and under the bottom vertical response curves (*Fig. 3.93*), showing a deeper null at 30° than at 15°. This is easier to see in the simultaneous

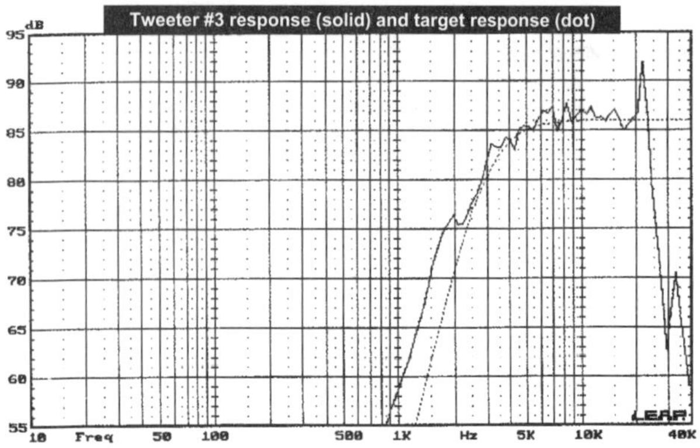

— Tw sect 3 + lpad p6
······· Generic Hp

FIGURE 3.61

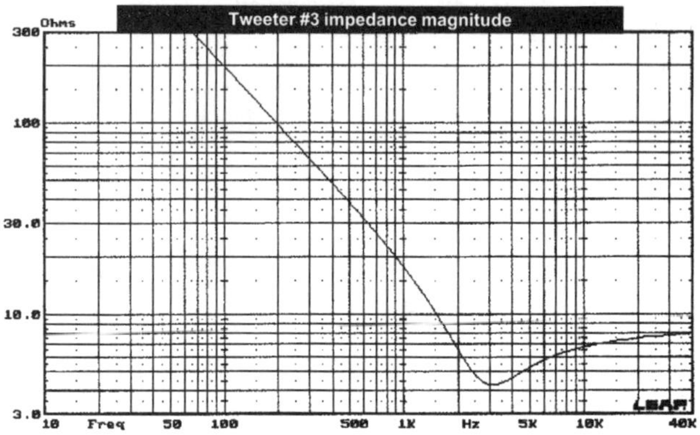

— Tw sect 3 + lpad p6

FIGURE 3.62

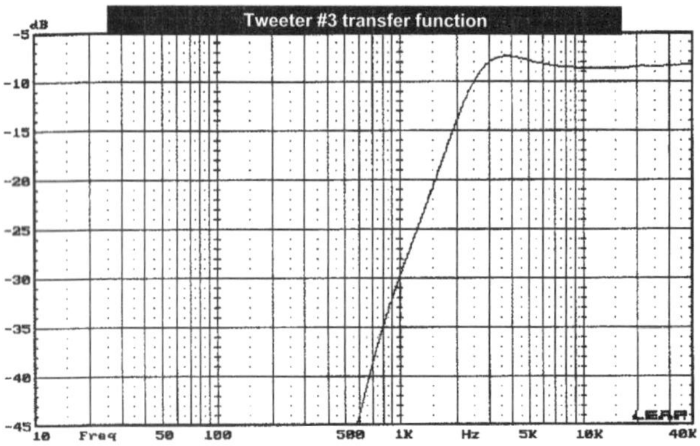

— Tw sect 3 + lpad p6

FIGURE 3.63

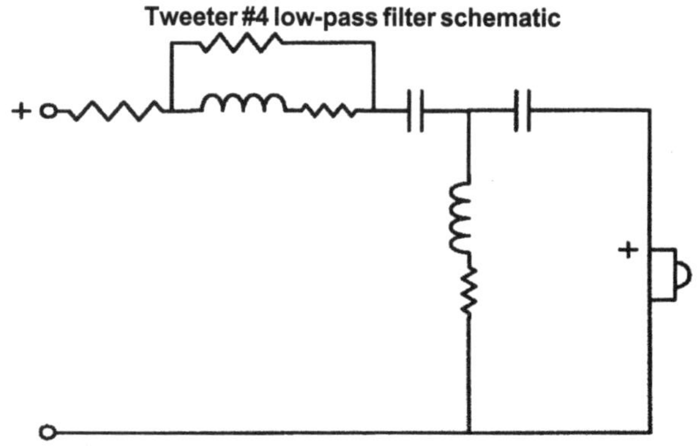

FIGURE 3.64

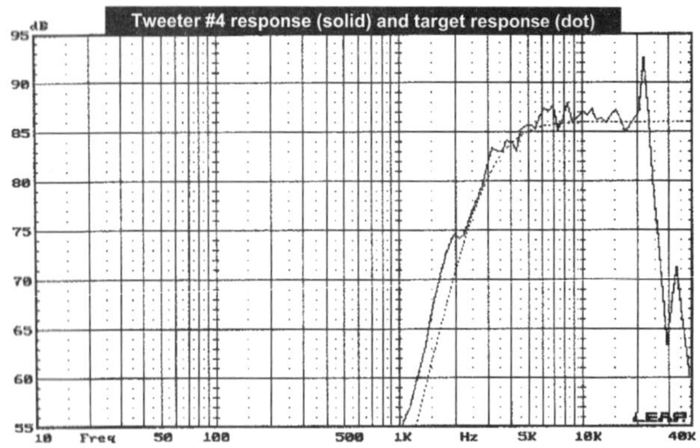

— Tw at 3000 with contour p17
······· Generic Hp

FIGURE 3.65

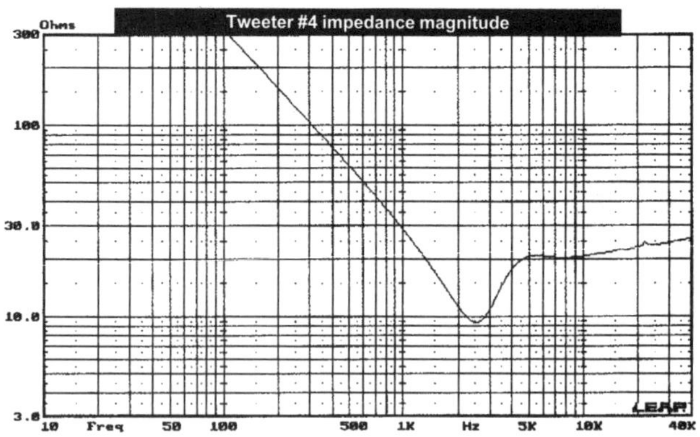

— Tw at 3000 with contour p17

FIGURE 3.66

comparison of each up/down vertical response curve at 15° in *Fig. 3.94* and 30° in *Fig. 3.95*. The obvious conclusion is that asymmetrical crossovers have asymmetrical lobing. This is the primary reason I mounted the woofer above the tweeter on this speaker's front panel. The arrangement results in the smoothest response from normal listening height downward.

An equally compelling argument could probably be made for placing the tweeter above and the woofer below. Either way, however, the "flaw" (if that is how it is interpreted) is not critical.

Left and right channels have well-matched response curves (*Fig. 3.96*). Using only two random driver samples, the curves are within

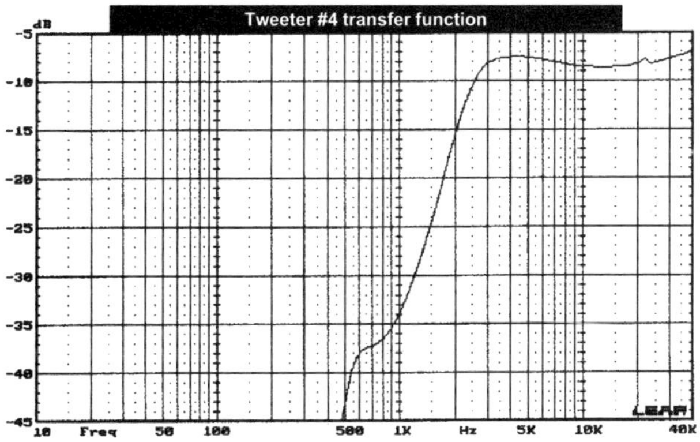

— Tw at 3000 with contour p17

FIGURE 3.67

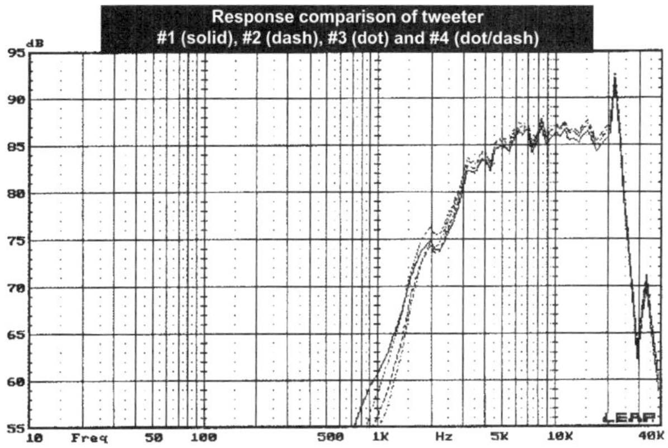

— Tw sect 4 + lpad p7
······ Tw sect 3 + lpad p6
- - - Tw sect 1 + lpad p5
-·-· Tw at 3000 with contour p17

FIGURE 3.68

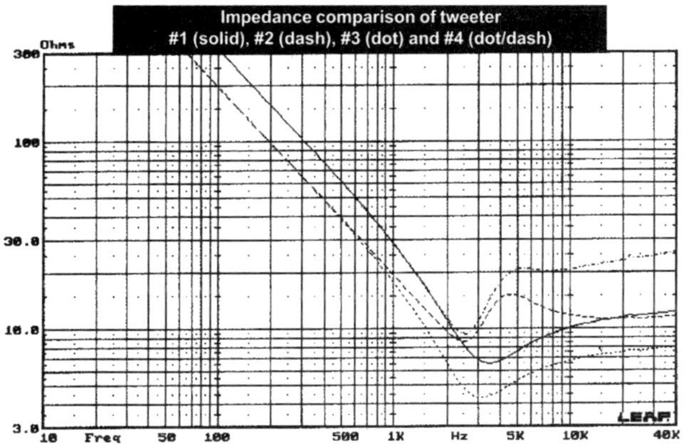

— Tw sect 4 + lpad p7
······ Tw sect 3 + lpad p6
- - - Tw sect 1 + lpad p5
-·-· Tw at 3000 with contour p17

FIGURE 3.69

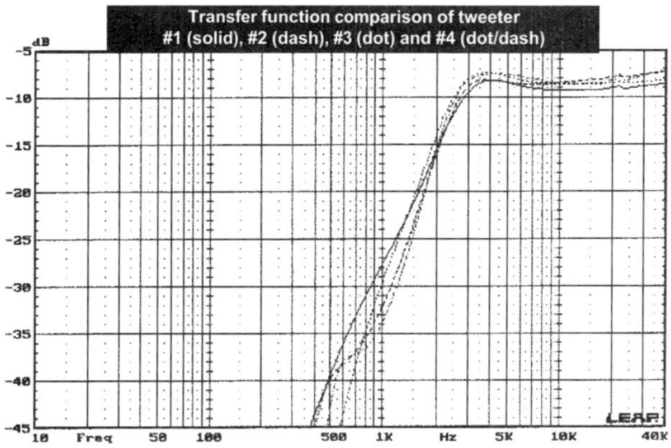

— Tw sect 4 + lpad p7
······ Tw sect 3 + lpad p6
- - - Tw sect 1 + lpad p5
-·-· Tw at 3000 with contour p17

FIGURE 3.70

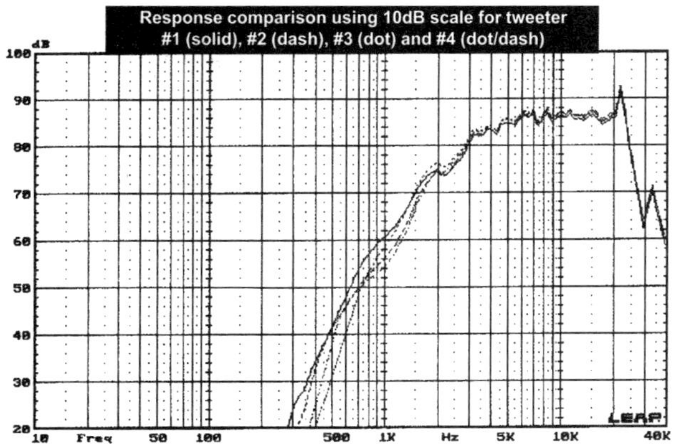

— Tw sect 4 + lpad p7
······ Tw sect 3 + lpad p6
- - - Tw sect 1 + lpad p5
-·-· Tw at 3000 with contour p17

FIGURE 3.71

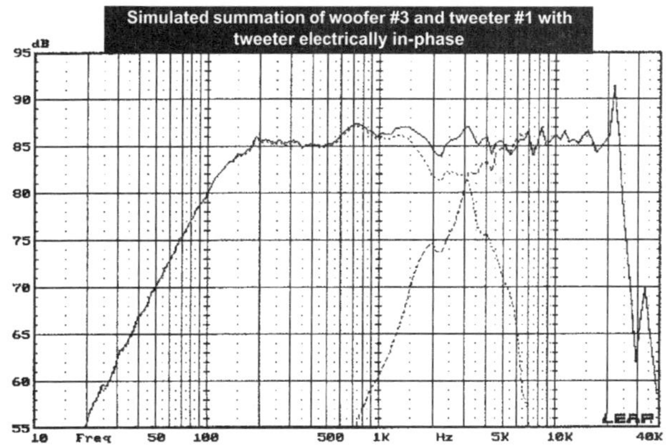

— Sum 13 in phase
······ Wof 3 at 2200 p20
- - - Tw sect 4 + lpad p7

FIGURE 3.72

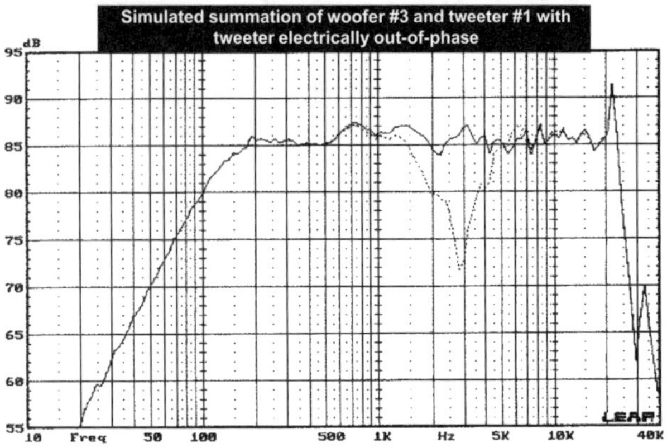

— Sum 13 in phase
········ Sum 13 out phase

FIGURE 3.73

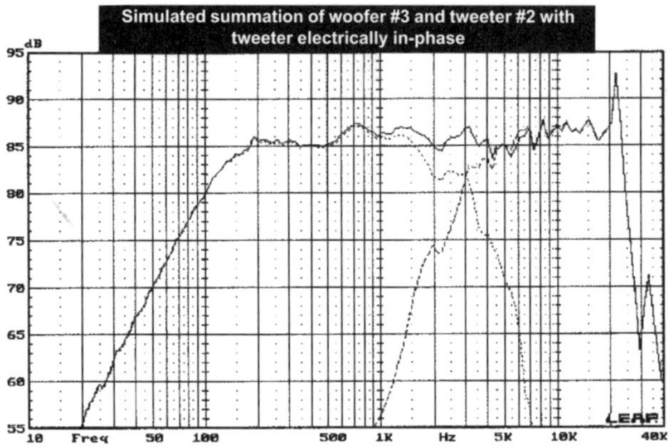

— Sum 11 in phase
········ Wof 3 at 2200 p20
- - - Tw sect 1 + lpad p5

FIGURE 3.74

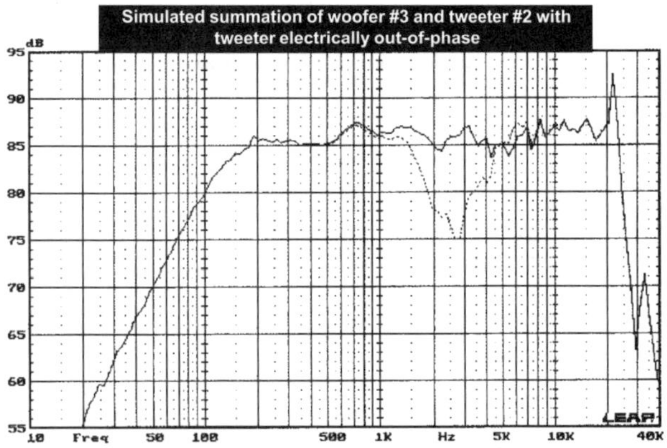

— Sum 11 in phase
········ Sum 11 out phase

FIGURE 3.75

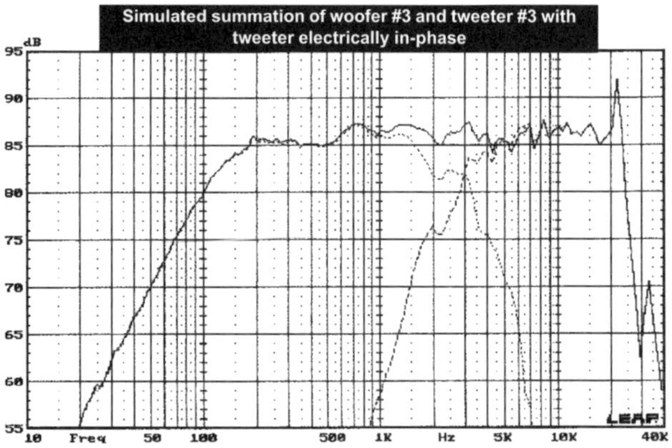

— Sum 12 in phase
········ Wof 3 at 2200 p20
- - - Tw sect 3 + lpad p6

FIGURE 3.76

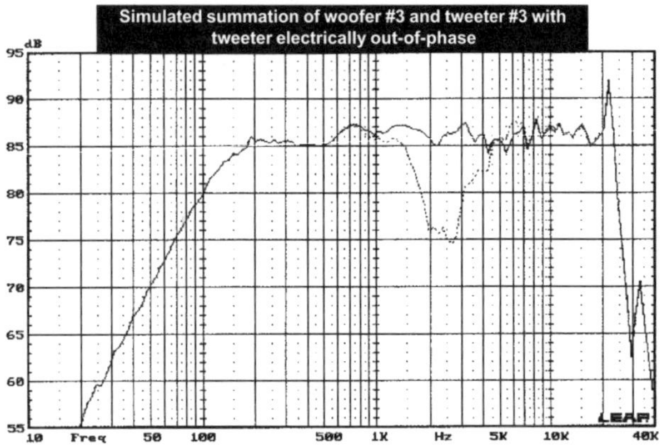

— Sum 12 in phase
········ Sum 12 out phase

FIGURE 3.77

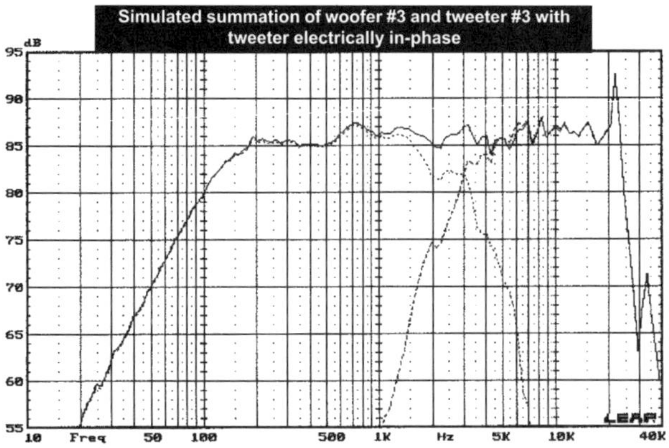

— Sum 14 in phase
········ Wof 3 at 2200 p20
- - - Tw at 3000 with contour p17

FIGURE 3.78

±0.5dB except for two small regions between 3-4kHz, and 12-14kHz, where the response is out 1dB.

Figure 3.97 compares the tweeter in-phase system response with the tweeter out-of-phase system response as well as both high-pass and low-

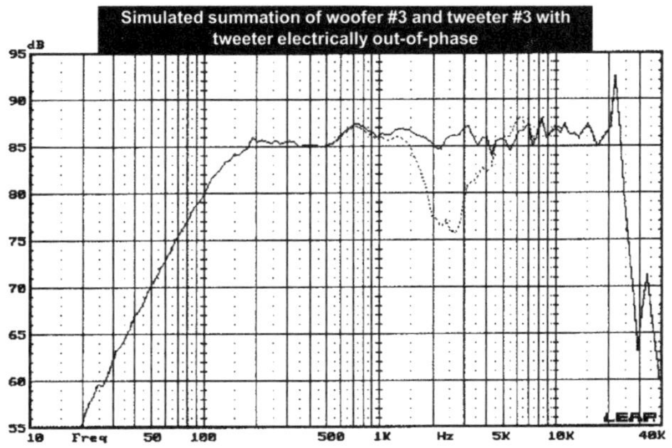

—— Sum 14 in phase
········ Sum 14 out phase

FIGURE 3.79

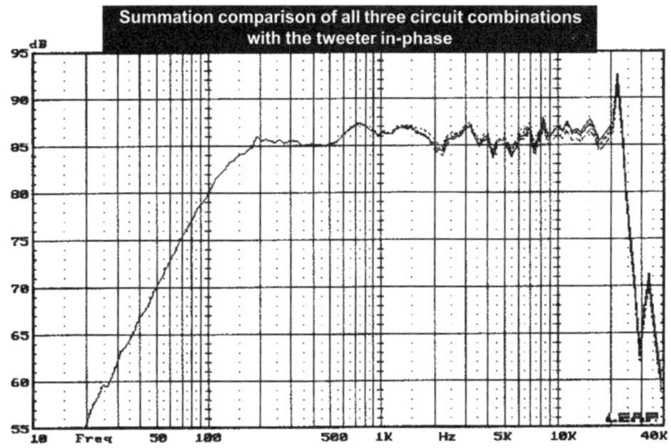

—— Sum 11 in phase
········ Sum 12 in phase
– – – Sum 13 in phase
–·–· Sum 14 in phase

FIGURE 3.80

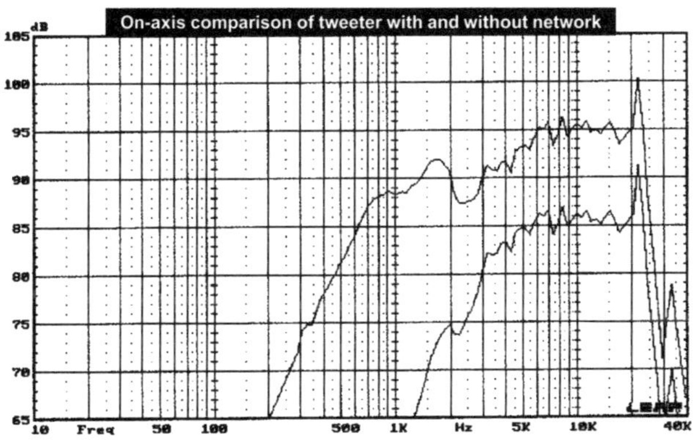

—— Focal T120ti
········ Tw sect 4 + lpad p7

FIGURE 3.81

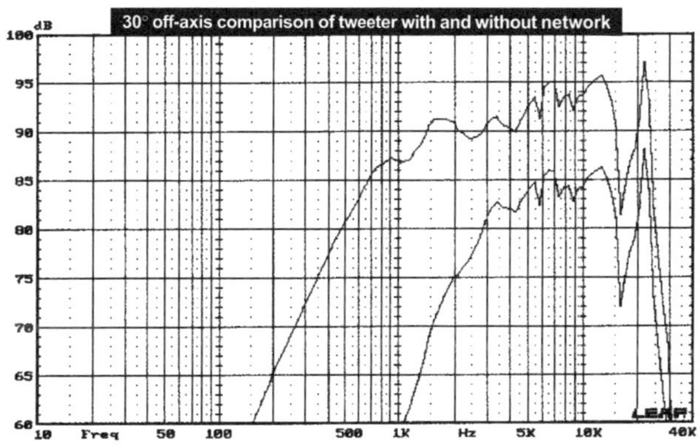

—— Focal T120ti
········ Tw sect 4 + lpad p7

FIGURE 3.82

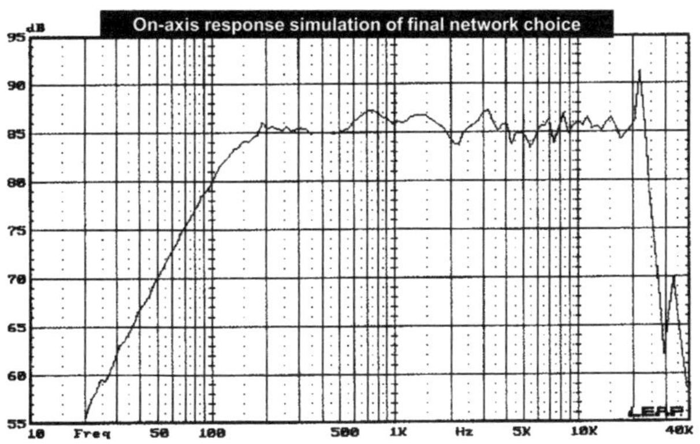

—— Wof 3 at 2200 p20
········ Tw sect 4 + lpad p7

FIGURE 3.83

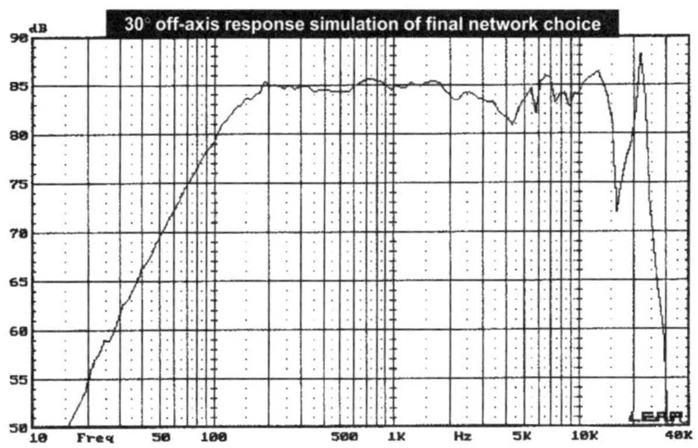

—— Wof 3 at 2200 p20
········ Tw sect 4 + lpad p7

FIGURE 3.84

pass responses. The out-of-phase null occurs somewhat below the 3kHz crossover frequency and the profile is asymmetrical, as we expected. The responses are primarily additive with only a small area between 5-6kHz where the phase is subtractive.

Figures. 3.98 and *3.99* show the impedance and Nyquist complex impedance plots. The low impedance point is 5Ω at 4kHz, while the maximum capacitive load is 35.3° at 2,072Hz indicating that this speaker should present no drive difficulties to any amplifier.

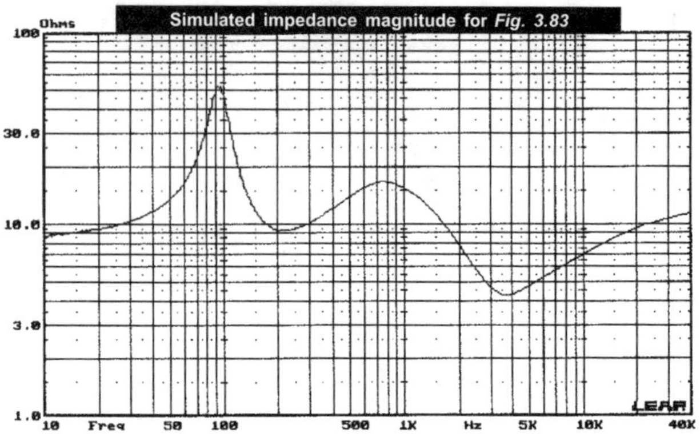

— Wof 3 at 2200 p20
······· Tw sect 4 + lpad p7

FIGURE 3.85

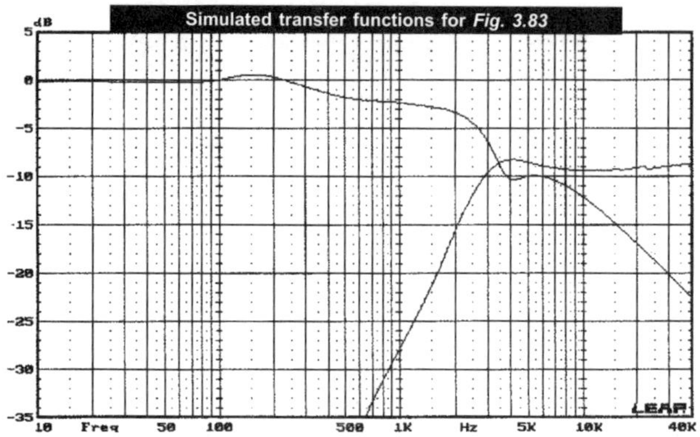

— Wof 3 at 2200 p20
······· Tw sect 4 + lpad p7

FIGURE 3.86

Final crossover schematic for Focal 7N515 and T120ti

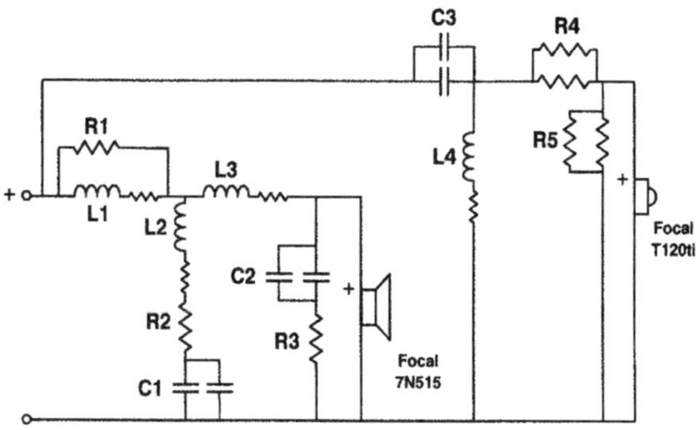

FIGURE 3.87

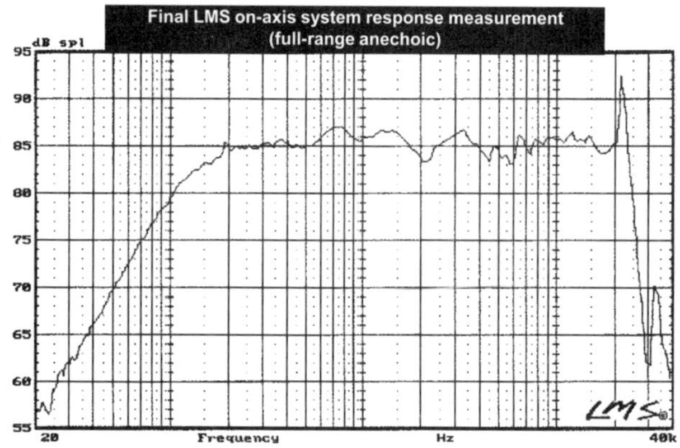

— Focal 7N515/T120ti

FIGURE 3.88

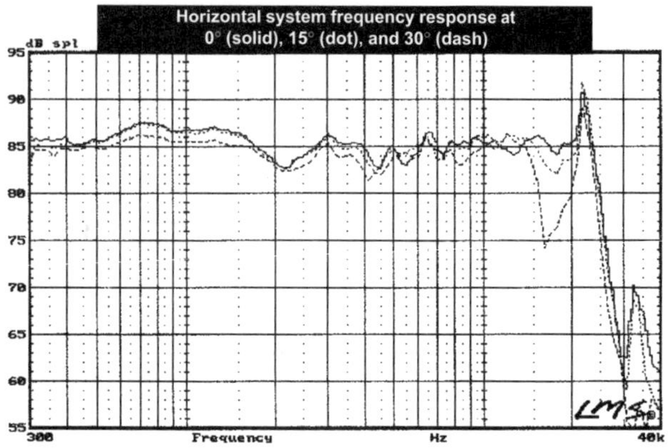

— Full range
······· Sys 15
- - - Sys 30

FIGURE 3.89

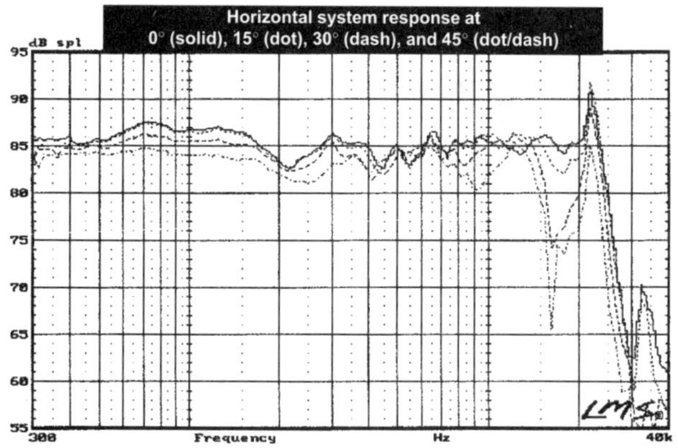

— Full range
······· Sys 15
- - - Sys 30
-·-· Sys 45

FIGURE 3.90

The last performance graph for the two-way satellite system is the MLSSA cumulative spectral decay curve in *Fig. 3.100*. The sharp resonance of the tweeter breakup mode is easy to see, but at a frequency of 25kHz it is doubtful whether this has an effect on sound quality. The midrange appears relatively uncluttered and settled to below -30dB by 1.5μs, which is considered good.

The total system response incorporating the subwoofer and the Marchand XM16 electronic crossover (shown in *Fig. 3.101*) is considerably extended and depicted in *Fig. 3.102*. With the

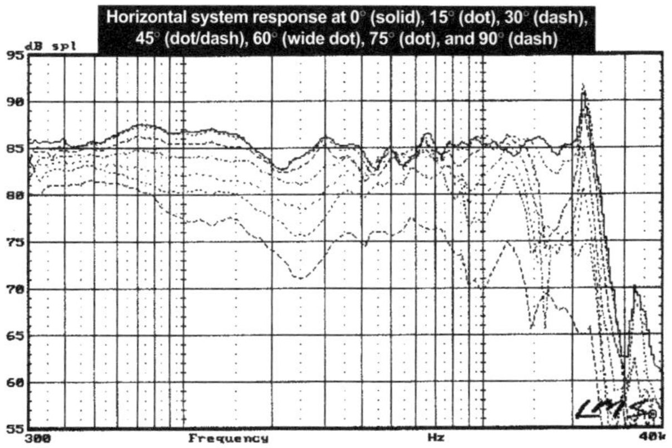

—— Full range
········ Sys 15
– – – Sys 30
–·–· Sys 45
· · · Sys 60
········ Sys 75
– – – Sys 90

FIGURE 3.91

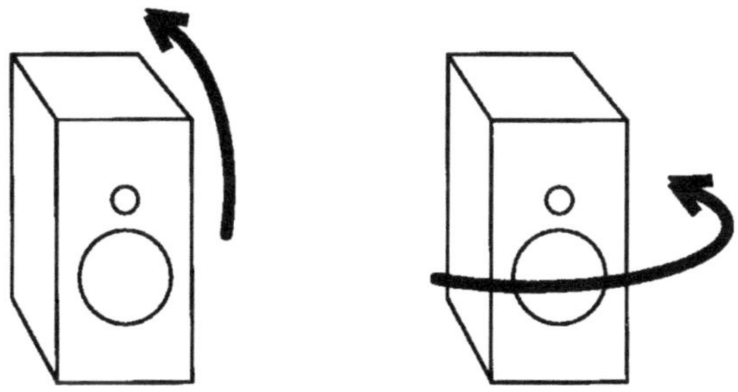

FIGURE 3.91B

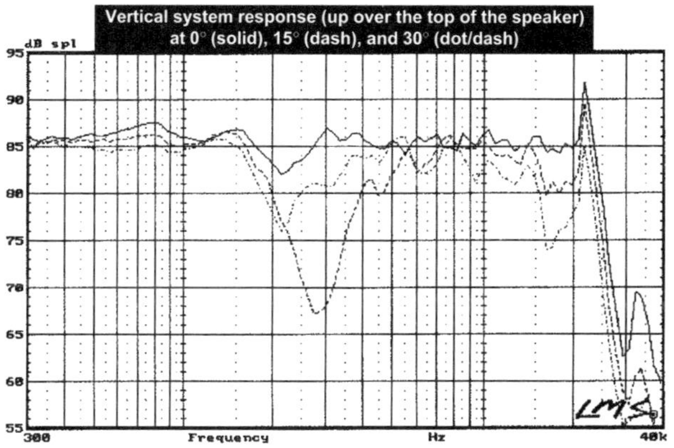

—— Sys 0 H
– – – H + 15
–·–· H + 30

FIGURE 3.92

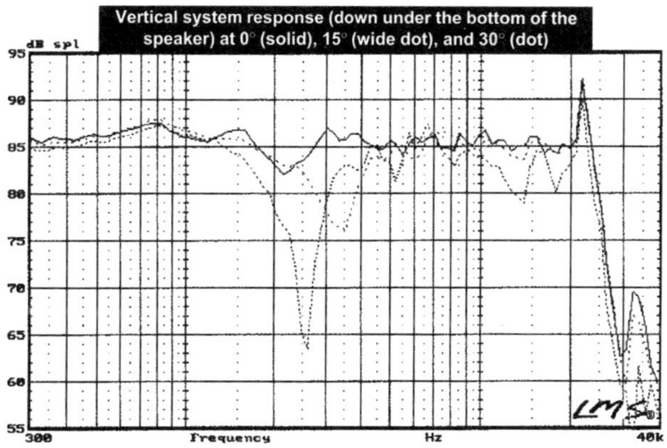

—— Sys 0 H
········ H - 30
· · · H - 15

FIGURE 3.93

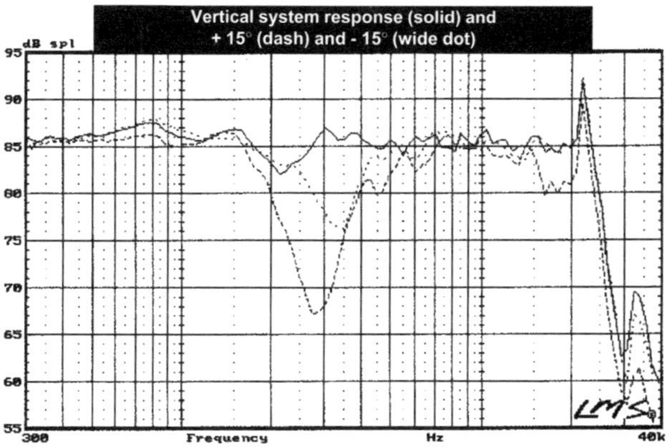

—— Sys 0 H
– – – H + 15
· · · H - 15

FIGURE 3.94

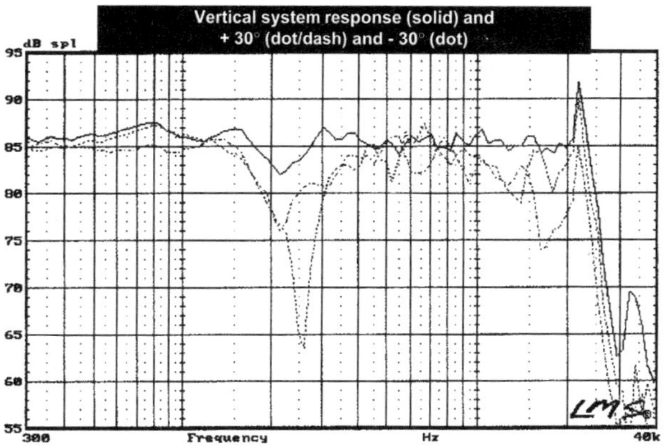

—— Sys 0 H
········ H - 30
–·–· H + 30

FIGURE 3.95

subwoofer, the total system response is ±3dB from 28Hz≈22kHz. The symmetry of the acoustic slopes using the fourth-order low-pass and second-order high-pass configuration with the Marchand XM16 is illustrated in *Fig. 3.103*. The crossover frequency came very close to the planned 120Hz

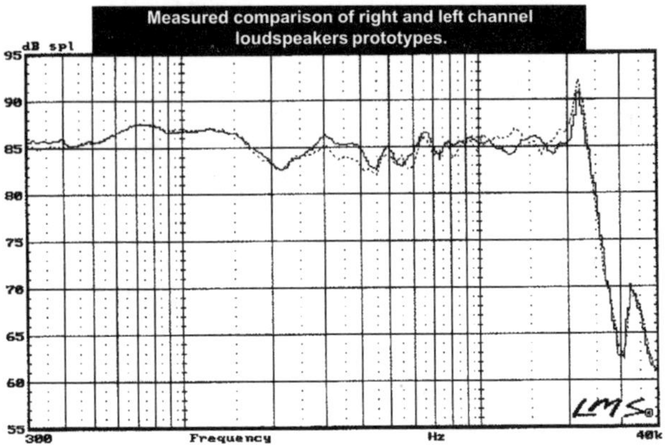

— Full range
······· Sys 0 b

FIGURE 3.96

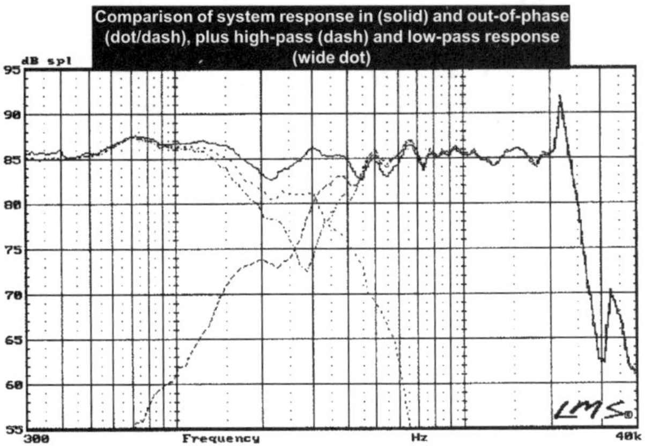

— Full range
-·-· Sys 0 rev
- - - Wof
--- Tw

FIGURE 3.97

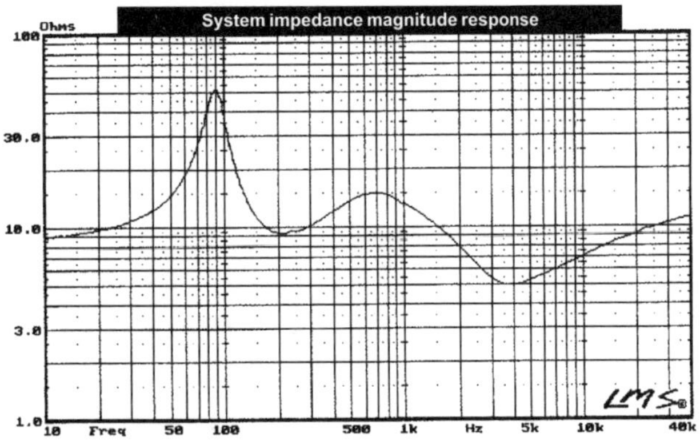

— M3 sys imp

FIGURE 3.98

Complex impedance curve (Nyquist plot)

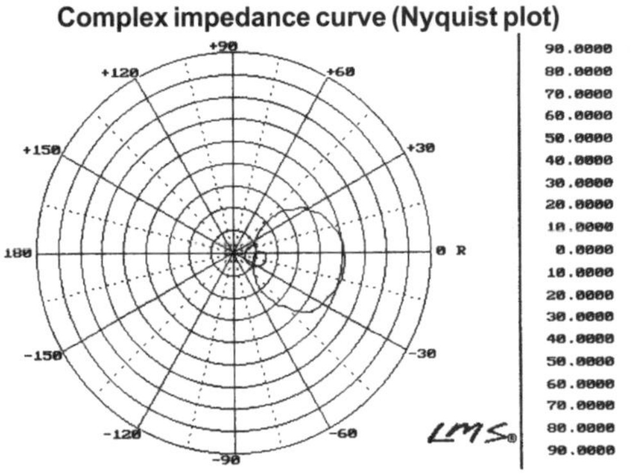

······· M3 sys imp

FIGURE 3.99

MLSSA cumulative spectral decay plot of Focal 7N515 and T120ti

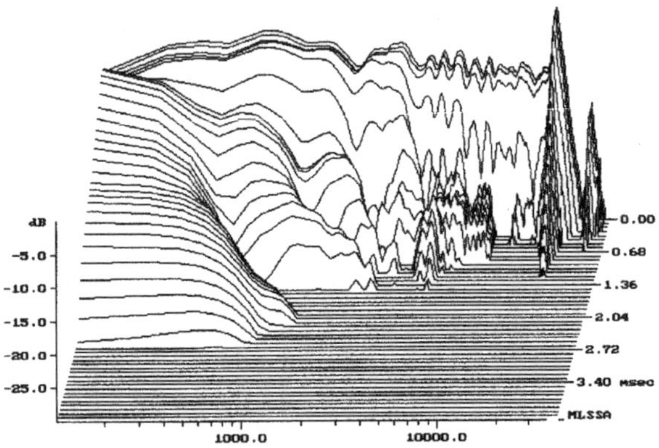

FIGURE 3.100

Photograph of Marchand Electronics XM16 electronic crossover

FIGURE 3.101

by combining the acoustic high-pass with an electronic high-pass.

3.100 Cabinet Performance

Constructing the satellite 7N515/T120ti speaker enclosure is somewhat different from the previous two-way speakers. The box is made from 0.75"

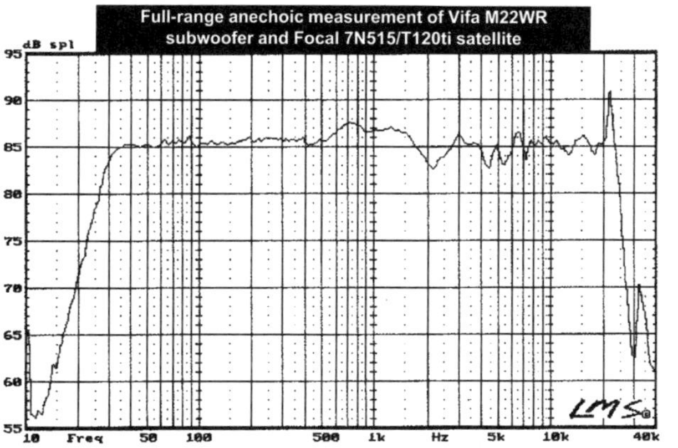

-·-· M2 + sub

FIGURE 3.102

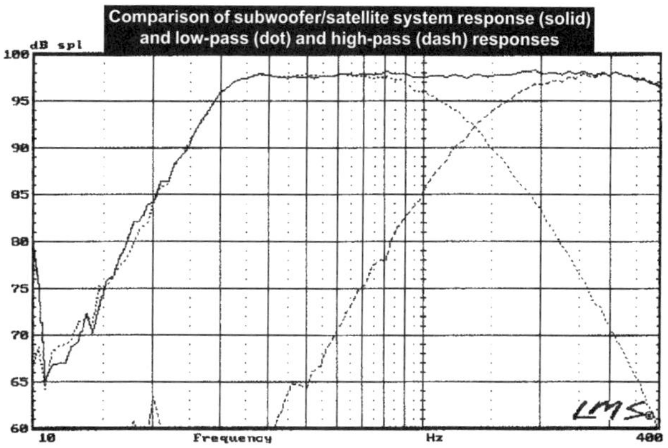

—— M4 sub/sat
········ M4 sub
- - - M4 sat

FIGURE 3.103

MLSSA accelerometer CSD plot of satellite box side panel

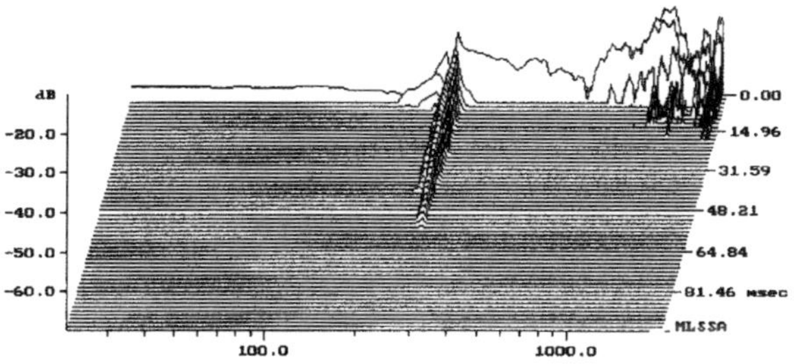

FIGURE 3.104

MLSSA accelerometer CSD plot of satellite box top panel

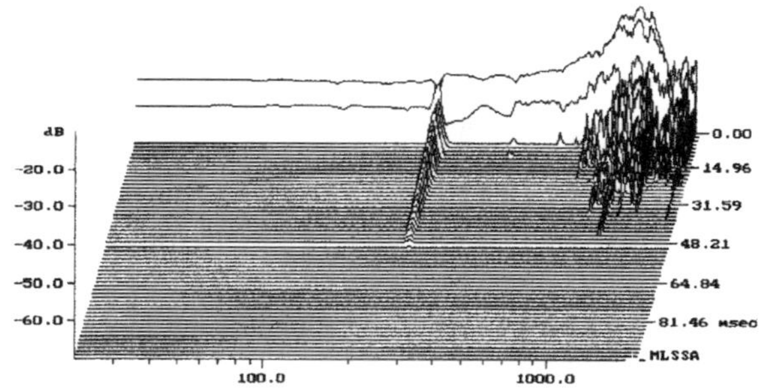

FIGURE 3.105

MDF instead of 1" MDF which results in a front baffle thickness of 1.5" rather than 1.75". Also, since the woofer cavity is so small I divided it with a solid 0.75" MDF panel. However, like the speaker in *Chapter 1*, I used an 8-inch diameter Sonotube to form a rather large cavity behind the woofer. This forms a large air space in the top part of the enclosure, which, as in the other speakers I filled this space with sand, produced a cabinet which weighed 35 lbs.

I constructed the subwoofer with 1" MDF using a vertical shelf brace and a horizontal crossbrace for damping (*Fig.3.113*).

Once again I did an accelerometer test. The MLSSA analyzer results are illustrated in *Figs. 3.104-3.106* for the two-way satellite and in *Figs. 3.107-3.109* for the subwoofer (both indicate the side, top, and rear accelerometer mounting locations). These plots can be compared to those in *Chapters 1* and *2*, and also show good box damping. The satellite's rear panel was somewhat more noisy (these are only relative measurements) than the other speakers. In the upper frequencies, the subwoofer rear panel, with only its single cross brace, was much noisier. Since the subwoofer will be rolled off at 100Hz, most of these modes will not be activated.

3.110 Construction Details

The two-way satellite construction details are given in *Fig. 3.110.* Construction is similar to the earlier projects, but I will review to include some of the details for this speaker. While the cabinet's damping characteristics are quite good, the design uses relatively simple construction techniques. The top, bottom, rear baffle and left and right walls are cut from 0.75" MDF, and assembled using glue and staples (nails). The four walls are butt-jointed for ease of assembly.

I limited internal reflections and standing wave modes by using an internal rear enclosure wall made from 8-inch diameter Sonotube. I sliced the 0.25-inch thick cardboard in half lengthwise, forming a semicylinder and cut to fit the inside height of the box. Position this on the enclosure's inside front and staple it to the side walls. I then

MLSSA accelerometer CSD plot of satellite box rear panel

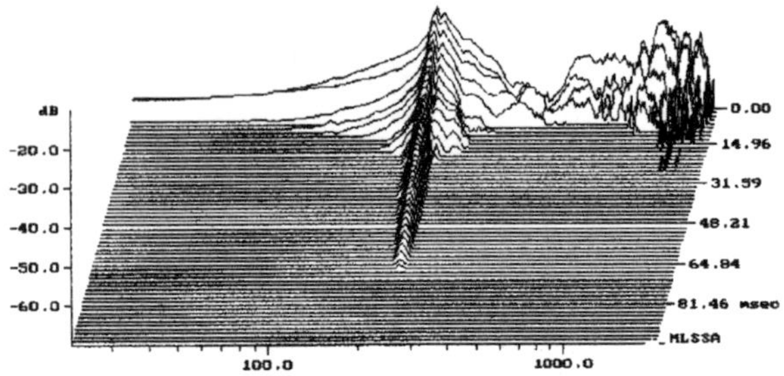

FIGURE 3.106

MLSSA accelerometer CSD plot of subwoofer box side panel

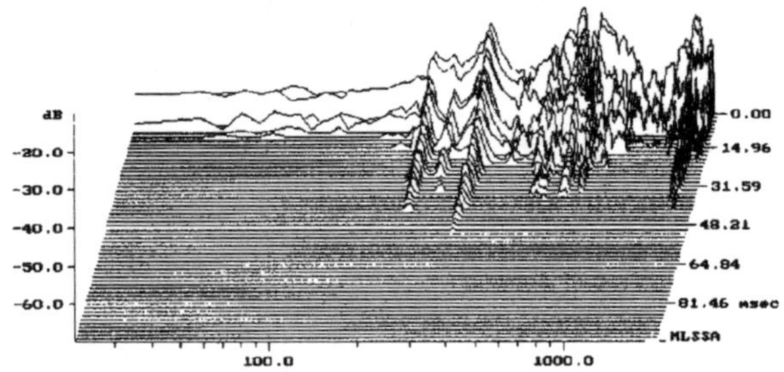

FIGURE 3.107

MLSSA accelerometer CSD plot of subwoofer box top panel

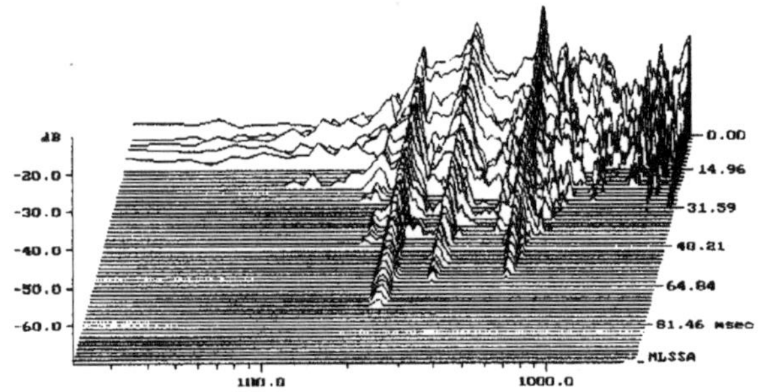

FIGURE 3.108

MLSSA accelerometer CSD plot of subwoofer box rear panel

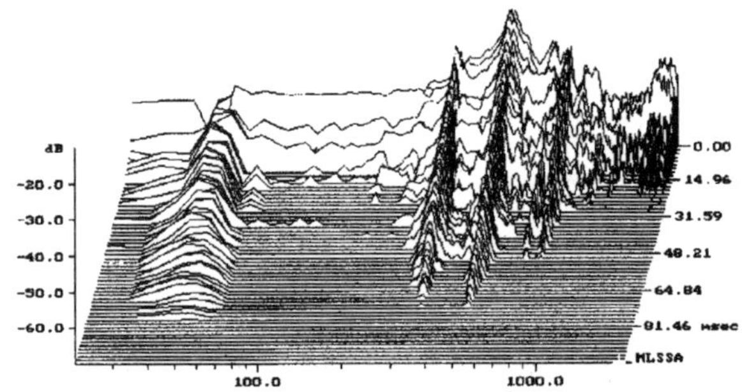

FIGURE 3.109

sealed the cardboard's edges with a combination of hot glue and silicone. After the cabinet is completely assembled, wired and finished, the space formed in the rear of the box is filled with sand through a 0.5" hole drilled at an angle into the internal baffle through the hole for the terminal cup. Locate this cup in the center of the enclosure's rear panel just below the woofer's internal baffle. Use a dowel plug to seal the holes for adding the sand.

The 1.5" front baffle consists of bottom and top layers of 0.75" MDF. Assemble it by first gluing and stapling the bottom layer to the box, and then glue and staple the top to the bottom. I routed the top baffle to mount the drivers flush with the baffle surface. To prevent the woofer from acoustically loading within the 1.5-inch thick panel, I routed the bottom layer with a curved 45° bit on the inner side of the woofer hole opening into the enclosure. The top layer has a 0.75" inset on the sides and top to accommodate the 0.75" oak quarter-round trim on the sides of the baffle and the angled trim piece at the top of the baffle (carefully measure the oak quarter-round because it varies considerably). *Figure 2.97* in *Chapter 2* illustrates the oak top trim piece, which extends across the top of the front baffle's top layer. The quarter-round is cut at the same angle as the oak side insets to produce a smooth angle across the top of the enclosure front.

After cabinet assembly, I painted the front baffle and enclosure bottom, and contact-cemented oak veneer to the sides, top and rear, and glued the oak trim in place.

To damp standing waves within the enclosure, I filled the inner space (not stuffed) with slightly teased Acousta-Stuf. If the density is increased, the Q will be lowered, and since this speaker already has a rather low Q_{TC} (measured at 0.69), avoid overstuffing. If you can, verify the speaker's Q_{TC}.

I mounted the woofer and tweeter using Well-Nut inserts. These have a nut embedded in a rubber insert and are frequently used to damp vibration from electric motors. They mount from the top of the hole and allow the driver to be removed easily. They also help isolate the drivers from the cabinet. I placed a tape foam gasket under the driver rims. The same type is used for weather stripping, although a superior neoprene rubber gasket is available from North Creek Music Systems.

As with the speakers in *Chapters 1* and *2* I did not locate the crossovers inside the enclosure, but inside the vertical stem of the stand. This not only isolates the components from vibration and influence of the driver magnetic fields, but also allows for a much larger component layout, allowing inductors to be placed further apart than would normally be possible. I installed two terminal sets, one each for the woofer and tweeter, on the the enclosure's rear panel. I mounted these in a small, (fits a 2 7/8" hole) round terminal cup from ACI less prone to vibrating, and fitted with a pair of gold binding posts. The additional pair, provided by Madisound, is identical to those from ACI. The four terminals connect directly to woofer and tweeter respectively (the new biwire cups would also work). The wires from the terminals is routed through a hole in the Sonotube, also sealed with silicone. Since this is a closed box, carefully seal all internal joints with a small bead of silicone to ensure integrity .

I made the grille frames from 0.25" MDF with 0.25" standoffs for the snap-type plastic fasteners. The standoffs are 0.5" x 0.5" pieces of 0.25" MDF glued to the frame. The standoffs are an attractive addition, and add the necessary thickness to mount the fastener. I attached black double-knit cloth to the frame with 3M spray fabric adhesive. This format works well, but you are free to use whatever method you prefer.

I used 0.75" MDF to construct the stands' bases and vertical stems detailed in *Fig. 3.111.* I finished the base and stem in black laminate with oak iron-on tape on the edges. Brass cone-type feet were mounted on the underside of the base by drilling shallow countersink holes. I joined the base to the stem with 1.5" black wood screws. Black Hole Pads cover the stem's inside walls to further reduce any vibration from the speaker cabinet.

Three pairs of gold binding posts, as well as the tweeter high-pass contour switch, can be mounted in a rectangular, blank terminal cup, inset into the stem. The terminal pairs handle input, high-pass to the tweeter, and low-pass for the woofer. Another pair could easily be added to provide biwire capability. Banana plug jumpers connect the terminals in the base to those in the speaker.

The stand's simple top is mounted to the stem using four wood screws. Adding rope caulk, the sort used to seal windows and doors, between the top and the vertical stem, further isolates the speaker from the stand. For additional isolation, I added rubber feet to the stand top, the type sold in electronic supply houses.

I chose 0.25" hardboard cut to fit the stem's inside side wall to mount the crossover components, with high- and low-pass sections on separate boards, although a single one would serve

Focal 7N515/T120ti box construction details

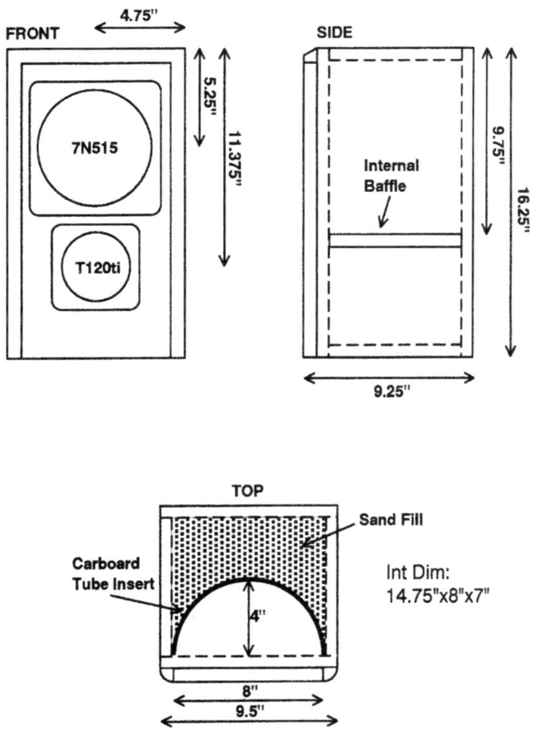

FIGURE 3.110

Construction detail of speaker stand

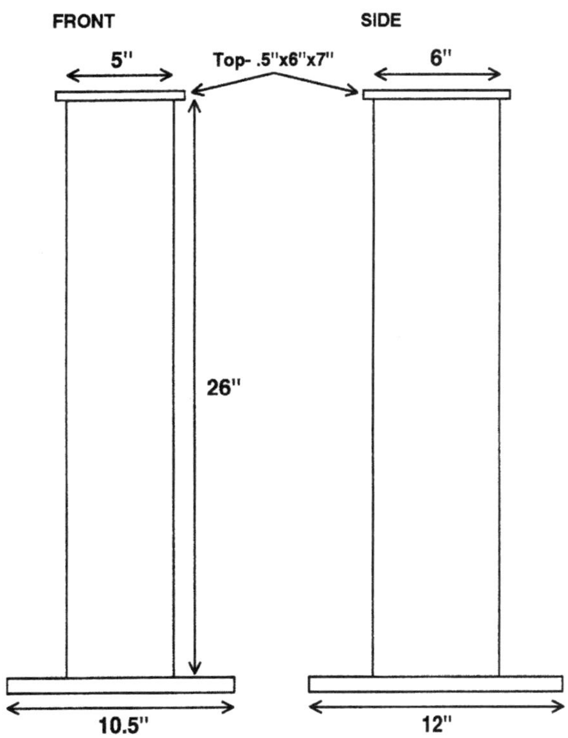

FIGURE 3.111

Photograph of crossover layout

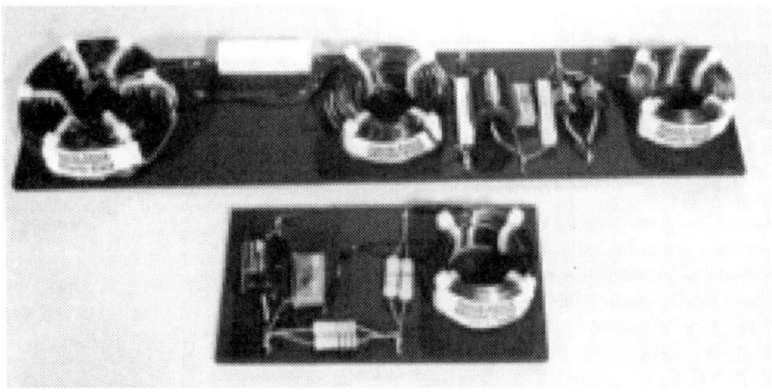

Photograph of Vifa M22WR box

FIGURE 3.112

Vifa M22WR box construction details

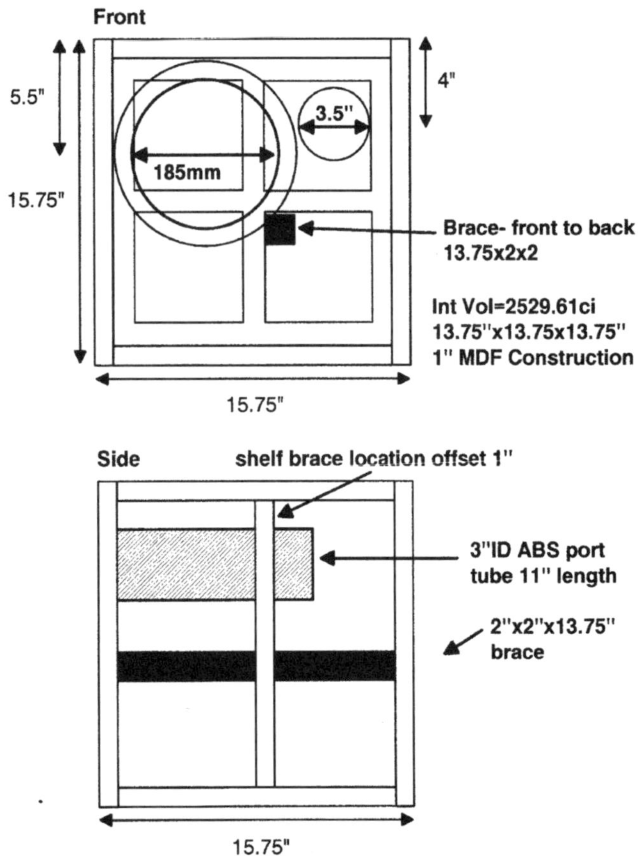

FIGURE 3.113

equally well. The stand's height makes it simple to spread components to provide the maximum inductor separation. Since this allowed me to mount the inductors some distance apart, I did not need to take the normal precaution of placing adjacent inductors at right angles and perpendicular to each other. Even though mounted in the same plane, inductive coupling did not appear to be a problem when I measured individual sections. However, you can provide about 10-20dB more attenuation by mounting them perpendicular at 90° angles using some kind of strapping to hold the inductors in place. Component layout is not critical except for inductor placement as shown in *Fig. 3.112*. Use silicone adhesive to affix the woofer section inside the stem's side wall. I mounted the tweeter section with the inductor at the top of the vertical stem in the same manner. I did all internal wiring in both the stands and the speaker cabinets with Monster cable for the woofers and Monster XP for the tweeters.

Brass split rivets arranged as layout points and mounted in drilled holes, served as connection points for soldering crossover components together. Silicone adhesive attached boards to the walls as well as inductors to the boards.

Construction details for the Vifa M22WR subwoofer are in *Fig. 3.113*. Constructed from 1" MDF, it includes shelf brace around the girth of the cabinet and a cross-brace joining front and back baffles. The terminal cup is located on the bottom of the enclosure, where brass cones provide room for the cable. Black Hole Pads on 50-60% of the interior surface, damps the box. I damped standing waves by lining the enclosure's three sides and back with R19 fiberglass, excepting the spot directly behind the port.

References

1. Mitchell, Peter, "The Ground Floor," *Stereophile*, Vol. 16, No. 3, March 1993

2. Hillerich, B.,"Acoustic Alignment of Loudspeaker Drivers by a Nonsymmetrical Crossover of Different Orders," *J*AES, Sept. 1989.

3. Hawkesford, M.O.J., "Asymmetric All-Pass Crossover Alignments," 93rd AES Convention, Preprint #3452.

Chapter Four

A DUAL WOOFER/ TWEETER TWO-WAY

4.10 Overview

The final design in this series combines two 6.5" woofers and a soft dome tweeter in what is often referred to as the "D' Appolito" configuration after loudspeaker designer and engineer, Joe D'Appolito. Mr. D'Appolito's original design, which started a huge fad in the industry, was based on the implementation of odd-order networks with the woofer-tweeter-woofer driver layout for the purpose of providing an unusually consistent vertical polar response, which it certainly did. The real benefit, however, is not so much the polar response using third-order filters as the fact that the configuration works extremely well for two-way loudspeakers regardless of the network type and provides a better image quality than a tweeter mounted on top of two woofers.

A majority of the W-T-W applications in the industry use fourth-order L-R networks rather than first- or third-order filters as did succeeding incarnations of Joe's own speakers. Regardless of the network type, the configuration provides the benefit of doubling the bass radiating area over a single small diameter driver, thereby increasing power handling and efficiency, plus retaining the advantage of using a small diameter driver to transition to a tweeter.

The drivers selected for this project are the Dynaudio 17W75 EXT woofer and the Morel MDT 33 soft dome tweeter.

4.20 The Woofers

The Dynaudio 17W75 EXT (extended range version) woofer is an excellent if rather unusual driver design. Constructed with a stamped steel metal frame which incorporates the magnet assembly housing as part of the magnetic circuit, the device locates the voice coil gap outside the perimeter of the magnet rather than in the center. This affords the use of a large 3" voice coil to drive the reverse conical poly cone at a point midway between its outer boundary and the apex. The result of all this manipulation is a small woofer with a very detailed upper midrange quality.

Following a 24-hour-break-in period, the samples were tested and the T/S parameters calculated as follows (the free-air impedance magnitude from which most of the parameters are derived is depicted in *Fig. 4.1*).

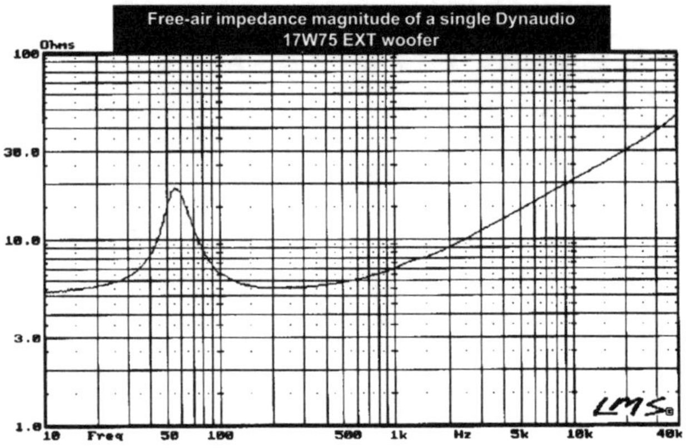

— Dyn17w afa

FIGURE 4.1

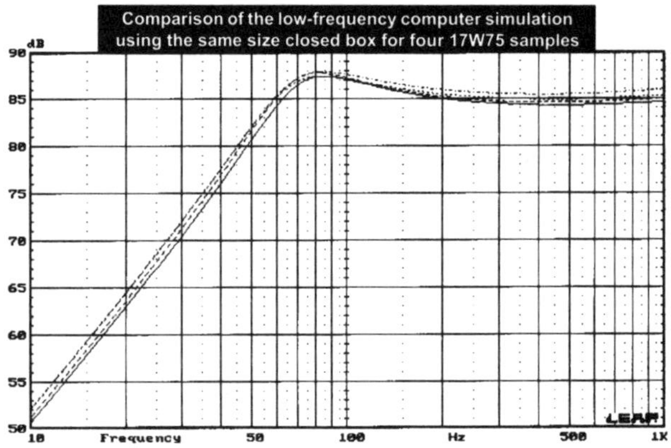

— Dyn 17w .8cf 2.83v b
······· Dyn 17w .8cf 2.83v
- - - Dyn 17w .8cf 2.83v c
-·-· Dyn 17w .8cf 2.83v d

FIGURE 4.2

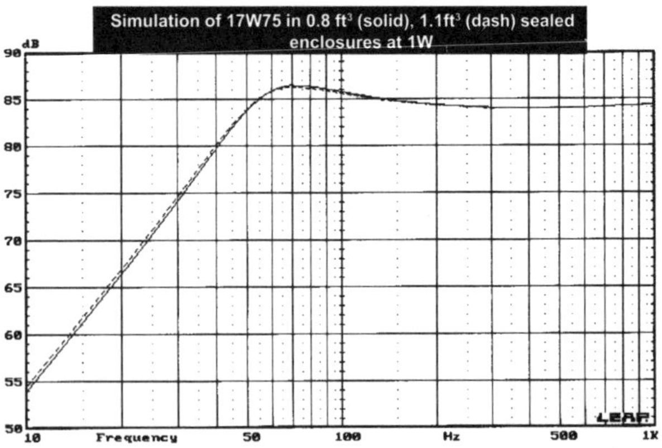

— Dyn 17w .8cf 2.83v
- - - Dyn 17w 1.11cf 2.83v

FIGURE 4.3

	A	B	C	D
Z_{NOM}	8	8	8	8
R_E (Ω)	5.3	5.3	5.2	5.0
S_d (M^2)	0.0133	0.0133	0.0133	0.0133
Bl (TM)	5.02	5.18	4.76	4.78
V_{AS} (ltr)	10	7.6	9.2	9.9
C_{MS} (μM/N)	400.4	302.6	369.1	396.4
f_S (Hz)	58.7	63.6	61.84	60.7
Q_{MS}	3.784	4.104	4.129	3.731
Q_{ES}	1.42	1.631	1.614	1.428
Q_{TS}	1.032	1.167	1.161	1.032
SPL (2.83V)	84.8dB	84.1dB	84.6dB	85.5dB
L_E(1kHz)	0.58	0.64	0.50	0.44

The largest variation among this set of samples was in the efficiency numbers. *Figure 4.2* shows the computer simulation in the same size enclosure for all four samples. The box Q and f_3 appear to be virtually identical and the only difference is about 1dB efficiency among the various drivers. By logically pairing the most with the least sensitive

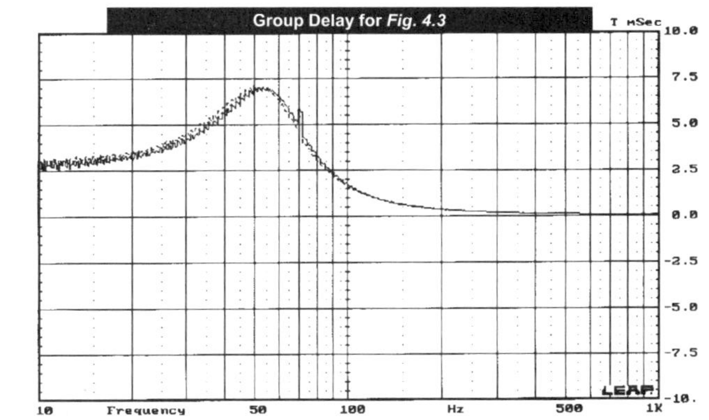

— Dyn 17w .8cf 2.83v
- - - Dyn 17w 1.11cf 2.83v

FIGURE 4.4

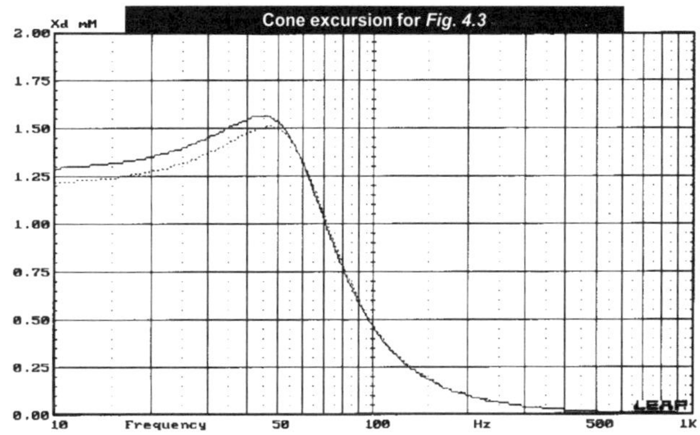

— Dyn 17w 1.11cf 2.83v
······· Dyn 17w .8cf 2.83v

FIGURE 4.5

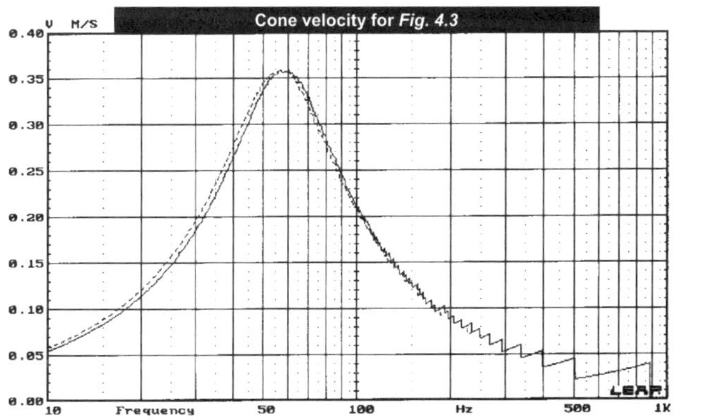

— Dyn 17w .8cf 2.83v
- - - Dyn 17w 1.11cf 2.83v

FIGURE 4.6

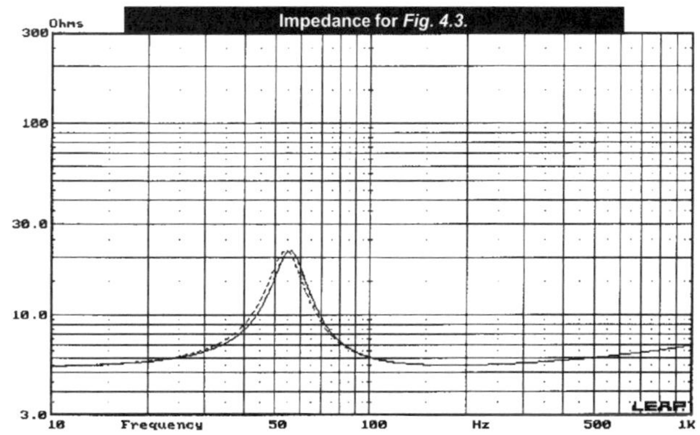

— Dyn 17w .8cf 2.83v
- - - Dyn 177w 1.11cf 2.83v

FIGURE 4.7

driver the finished prototype's left and right channel speakers had nearly identical curves, so this is of trivial concern.

The enclosure design is more or less a copy of the Dynaudio TWYNN speaker which uses two 17W75s and a D28 Dynaudio soft dome. This box design uses two separate sub-enclosures divided by an internal angled baffle with a Dynaudio Variovent for each cavity. Volume of the top speaker is 0.8 ft^3, while the volume of the lower woofer is 1.1 ft^3, creating what Dynaudio refers to as a "staggered" resonance system.

LEAP cannot model this type of "aperiodic" enclosure. However, as I demonstrated in Section 2.130 of the Fourth Edition of *The Loudspeaker Design Cookbook*, the Variovent type of loading does not alter the Q_{TC} and final f_3 in a way that cannot also be achieved using different box stuffing procedures. Knowing this, simulating this type of speaker loading was only a matter of increasing the percentage of fiberglass fill in the computer simulation to approximate the Variovent effect.

Figures 4.3-4.7 depict the 1W low-end computer simulation of the woofer in 0.8 ft^3 and 1.1 ft^3 enclosures for the woofers' frequency response, group delay, cone excursion, cone velocity and impedance magnitude. The first thing you notice is that there really isn't much difference in putting this woofer in different sized enclosures, which is due mostly to its high Q_{TS}. The second item which stands out is the apparent high box Q_{TC} for these woofers in this size enclosure. f_3 is a respectable 49Hz (for sealed box 6.5" drivers), but the phase angle at the -3dB point is 105° which coincides with a Q_{TC} of about 1.2. This is not surprising considering the shape of the "knee" in the group delay curve in *Fig. 4.4*.

Figures 4.8-4.11 show the woofers with the input raised to 5.2W illustrating the response, group delay, cone excursion, and cone velocity changes. Cone excursion at 5.2W is approximately 3.1-3.3mm, which is at the limits of linearity for the 17W75 which only has 2.75mm of cone travel available in one direction (X_{MAX} + 15% = 3.16mm). The SPL at this input level is

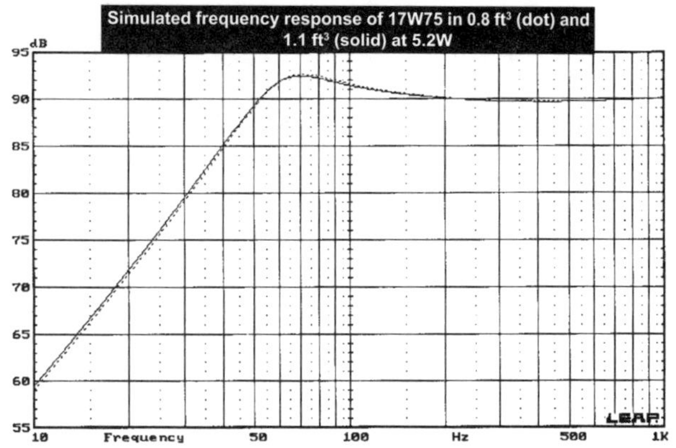

— Dyn 17w 1.11cf6.5v
.... Dyn 17w .8cf6.5v

FIGURE 4.8

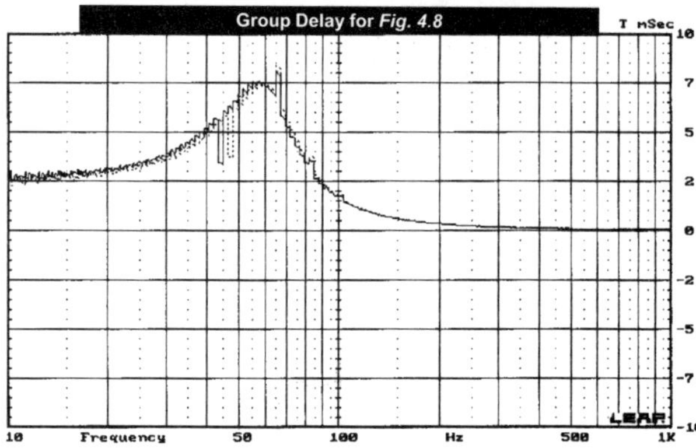

— Dyn 17w 1.11cf6.5v
.... Dyn 17w .8cf6.5v

FIGURE 4.9

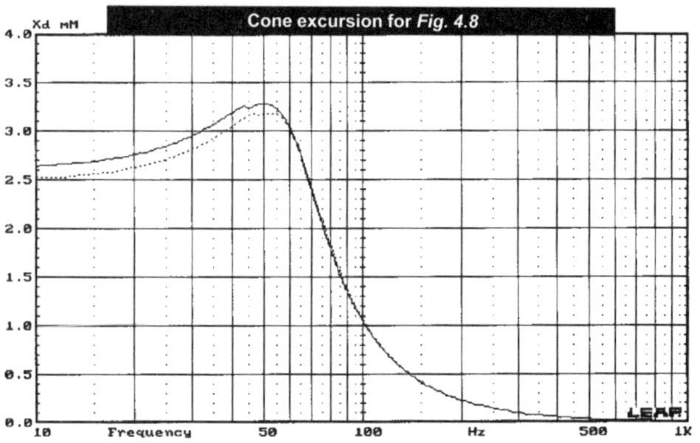

— Dyn 17w 1.11cf6.5v
.... Dyn 17w .8cf6.5v

FIGURE 4.10

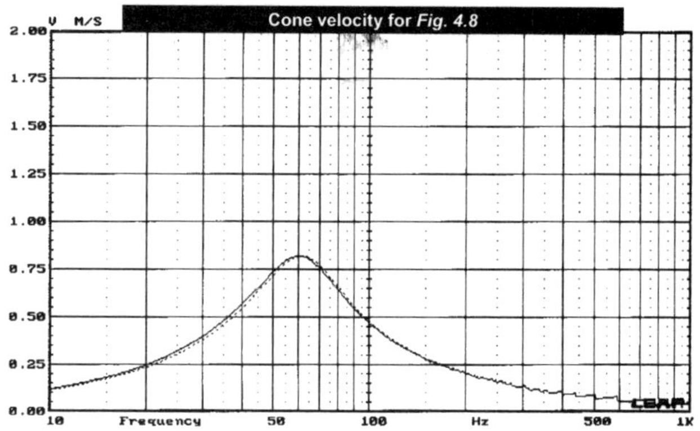

— Dyn 17w 1.11cf6.5v
.... Dyn 17w .8cf6.5v

FIGURE 4.11

only 90dB, so we can only expect an SPL of perhaps 96dB using two drivers wired in parallel, before the onset of audible distortion. Thus the dynamic range of these woofers is somewhat limited in terms of very high SPLs. Considering this driver's excellent midrange quality, it underlines the desirability of using a subwoofer.

I verified the single driver simulation by comparing an outdoor groundplane measurement of the prototype box (lined with Owens Corning Type 703 acoustic fiberglass--the stiff yellow type of fiberglass that weighs 3 lbs./ft² and Variovents installed) and scaling it to a 1W anechoic simulation as shown in *Fig. 4.12.* Note the SPL near the low-frequency roll-off is about 80dB (the 84dB efficiency numbers are calculated based on a half-space domain). If I lower the box Q_{TC} to 0.8 or 0.9 which is exactly what I did, and contour the response to be flat from the low-end roll-off of the speaker by using two woofers in parallel, the final efficiency will likely be somewhere between 84-85dB.

Q_{TC} of the enclosure and driver combination was lowered from 1.2 to 0.9 by stuffing the entire inner volume with R19 pink fiberglass (in addition to the 1.5" lining of 703 fiberglass) except for the area between the rear of the woofer and the Variovent. The latter I filled with a teased quantity of Mahogany Sound's Acousta-Stuf. I verified the Q_{TC} by measuring the impedance with LMS and exporting the file into LEAP, using the parameter measurement routine to calculate the final Q. Before the additional fiberglass and Acousta-Stuf treatment, the Q measured 1.27, but dropped to a more usable 0.93 afterward. *Figure 4.13* shows the comparison of the low-frequency groundplane response before and after I added damping. Note also that the excursion of the driver only gets worse as Q is lowered, although only marginally so in this case.

The full-range anechoic measurement of the two woofers is given in *Fig. 4.14* (with a measurement axis between the two drivers). Two individual samples are compared in *Fig. 4.15*

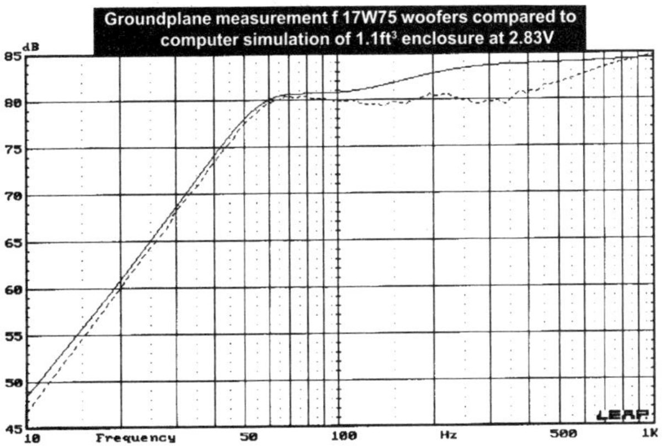

— Dyn 17w 1.11cf 2.83v
--- 17w groundplane 703+variovent

FIGURE 4.12

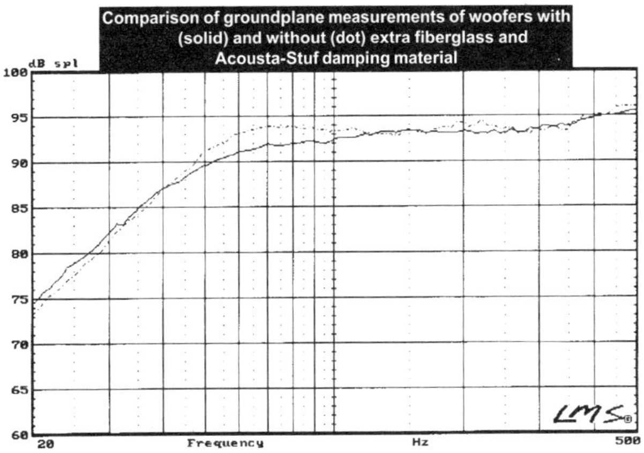

-·-· Qp no xover
— Qp r19+as

FIGURE 4.13

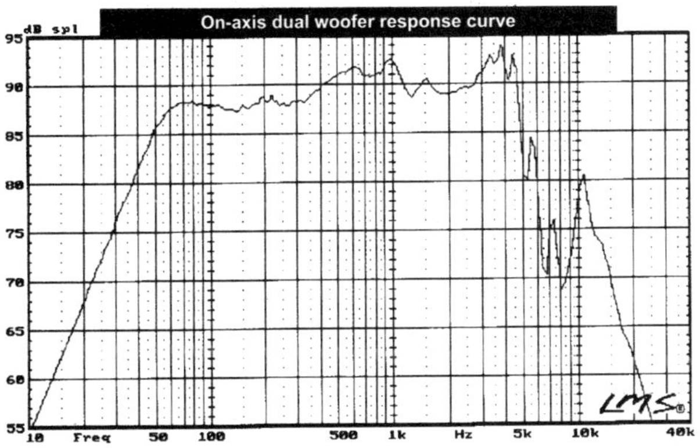

— Dyn 17W75 FR

FIGURE 4.14

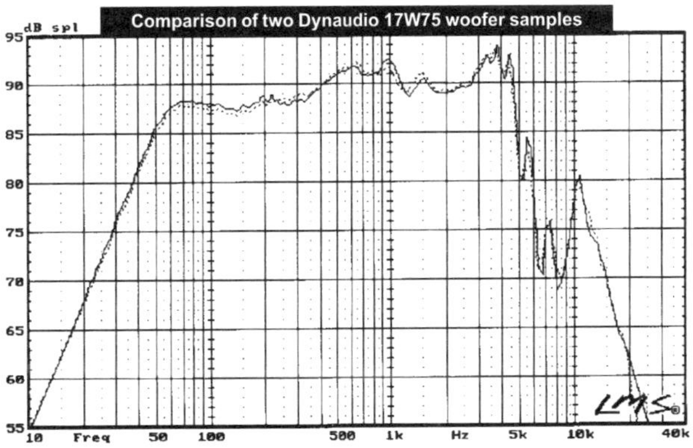

— Dyn 17W75 A
······· Dyn 17W75 B

FIGURE 4.15

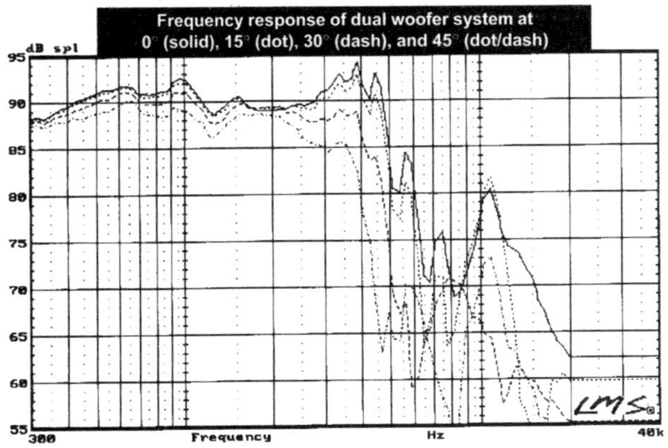

— Dyn 17W75 0
······· Dyn 17W75 15
--- Dyn 17W75 30
-·-· Dyn 17W75 45

FIGURE 4.16

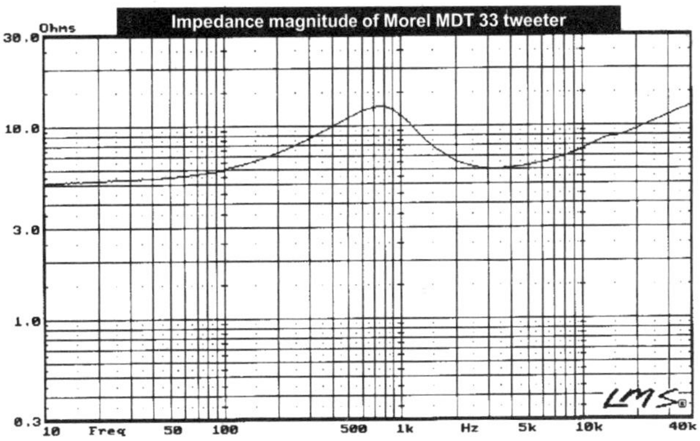

-·-· Mdt33a

FIGURE 4.17

(measured with the box Q_{TC} of 1.2 prior to adding damping material). As with most small diameter woofers, the response starts rising at about 200Hz and is up 4dB at 1kHz. In terms of quality control, the response of both samples was well within 0.5dB. The frequency response curves for 0°, 15°, 30°, and 45° are displayed in *Fig. 4.16* and demonstrate the excellent off-axis behavior of the

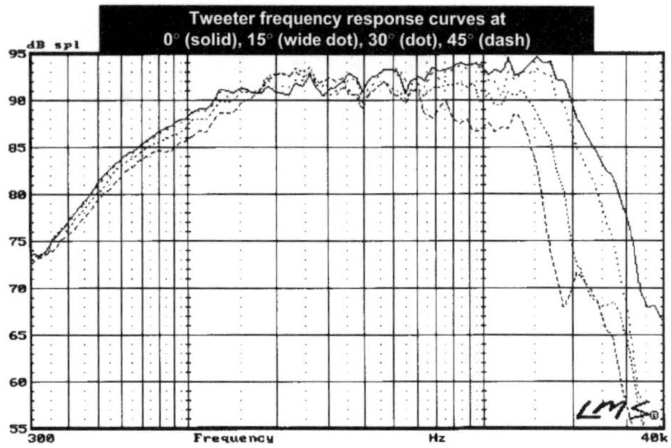

- - - Morel MDT33 15
—— Morel MDT33 0
.... Morel MDT33 30
- - - Morel MDT33 45

FIGURE 4.18

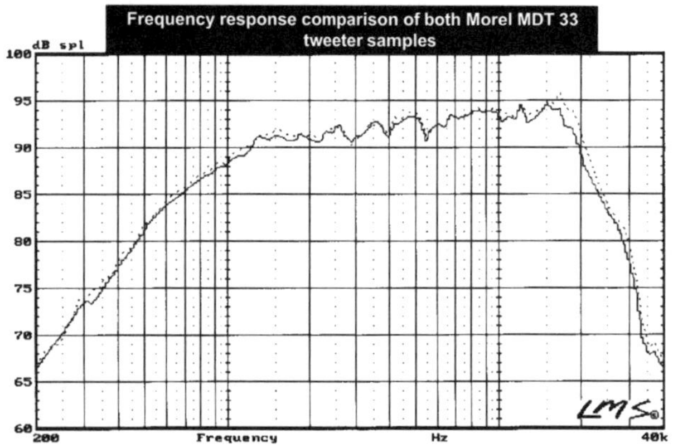

- - - Morel MDT33 B
—— Morel MDT33 A

FIGURE 4.19

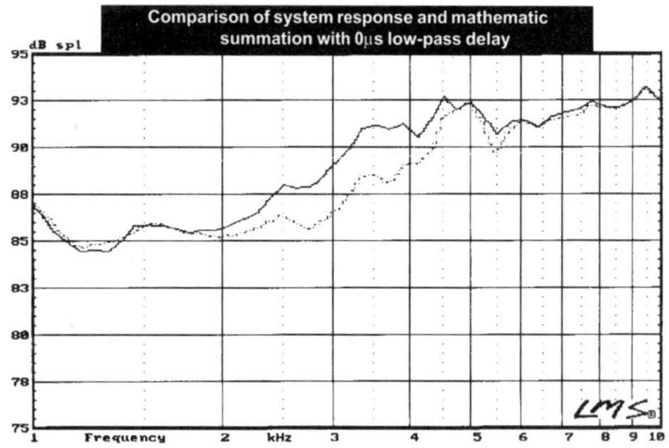

-·-· Sum 0 delay
—— Sys 0 net

FIGURE 4.20

17W75. As in the three 6.5" woofers discussed in *Chapters 1-3*, the off-axis response suggests a crossover frequency between 2-2.5kHz would provide the best system off-axis response (the 17W75 is down 6dB at 45° at 3kHz, and only 1dB at 2kHz).

4.30 The Tweeter

Morel's MDT 33 is an excellent example of the best in soft dome tweeter construction. It has a double magnet system, a vented pole piece and rear cavity, and an aluminum wire voice coil with magnetic fluid for additional cooling. Its 28mm hand-treated fabric dome is similar to Scan-Speak's D2905. With a low resonance of 700Hz, Morel specs the frequency range at 1.8kHz-20kHz, and supplies the MDT 33 in pairs matched to within 0.6dB. Specifications for the tweeter were derived from the impedance curve shown in *Fig. 4.17* and are as follows:

Z_{NOM}	8Ω
R_E	5.17Ω
f_S	694.6Hz
Q_{MS}	0.870
Q_{ES}	0.540
Q_{TS}	0.333
L_E (1kHz)	0.2mH
(20kHz)	0.04mH

The on- and off-axis response at 0°, 15°, 30°, and 45° is depicted in *Fig. 4.18*. On-axis response has a slight rise, but is nearly flat and contains no surprises. Off-axis behavior is also quite good, and typical of a high quality 1" cloth dome in the upper frequencies. As Morel promised, the two samples were well matched to within the specified 0.6dB, except for a small region above 17kHz which is evident in the response comparison in *Fig. 4.19*.

4.40 The Crossover Configuration

The interdriver time delay, which for all practical purposes is the physical distance between the centers of the two driver frontplates, is very small for the Dynaudio 17W75 and the Morel MDT 33. The Dynaudio woofer has a very shallow profile and has a frame designed to mount flat on a baffle surface instead of being inset (its contoured lip slopes up smoothly from the baffle at the basket rim thus avoiding problems). Measured from the front baffle, the frontplate center is only 1" behind the baffle surface. The Morel MDT 33's frontplate is set in from its faceplate about 0.3", farther than most tweeters I have used. This is ideal for minimizing the time delay, simply because of the way the two drivers are made.

The distance between them is only about 0.7", or a mere 50µs of delay. I confirmed this using the same technique as before, shown in *Figs. 4.20-4.23*, which compares the measured system response to the mathematically summed

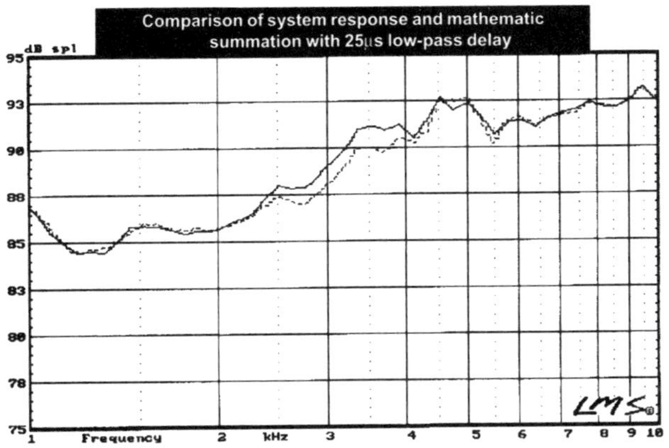

— Sys 0 net
- - - Sum 25us delay

FIGURE 4.21

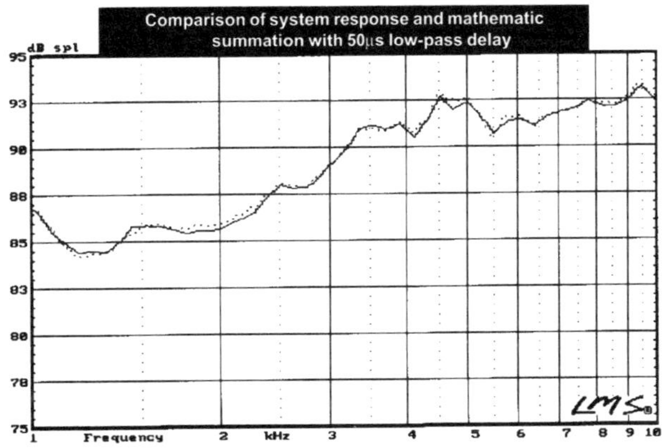

· · · Sum 50us delay
— Sys 0 net

FIGURE 4.22

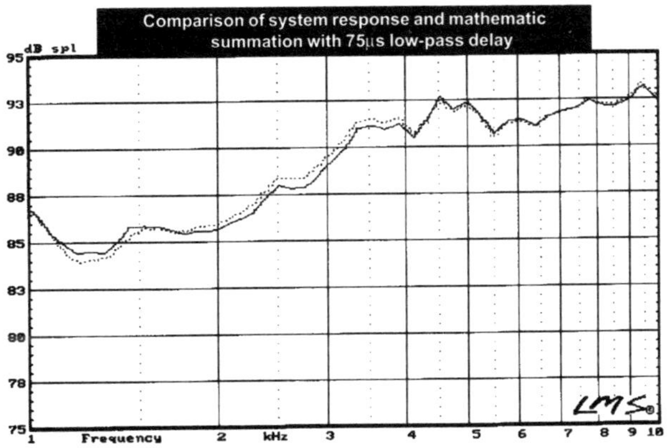

— Sys 0 net
······· Sum 75us delay

FIGURE 4.23

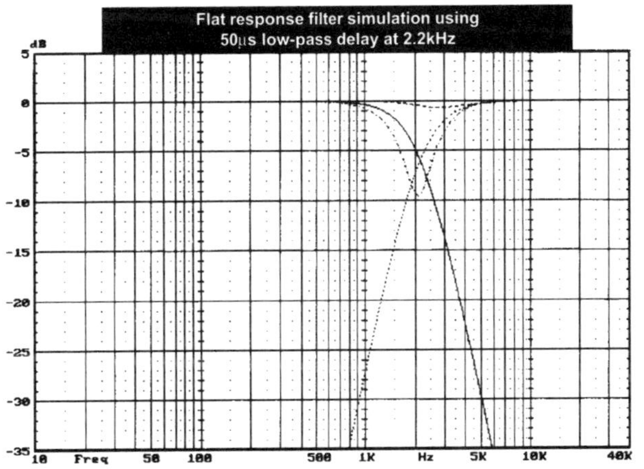

— Lo Pass Section Only: Magnitude
······· Hi Pass Section Only: Magnitude
- - - Sum=Lo+Hi (InPhase): Magnitude
-·-· Sum=Lo-Hi (Reverse): Magnitude

Lo Filtr: 4th-LnkRly		Hi Filtr: 4th-LnkRly	
Freq =	2200.00 Hz	Freq=	2200.00 Hz
Level=	0.00 dB	Level=	0.00 dB
Delay=	50.00µs	Delay=	0.00 µs

FIGURE 4.24

response with manually iterated amounts of added low-pass delay. Using a fourth-order L-R crossover at 2.2kHz, 50µs produces a 0.75dB dip just above the cross point (*Fig. 4.24*), which is perfectly acceptable. This was the easiest solution to the interdriver time delay problem of any of the four designs. Do nothing, it's close enough!

4.50 The Woofer Crossover

The full-range anechoic response of the woofer depicted in *Fig. 4.24* shows the Dynaudio 17W75 pair to be rising from approximately 200Hz and is +4dB higher at 1kHz. The response in this region is fairly smooth and the peak at 4kHz is probably at a frequency high enough not to be a problem. It becomes "buried" in the crossover stopband since it is nearly an octave above the crossover frequency.

I began the crossover design process by looking at the values generated by the LEAP Conjugate Designer to produce a flat impedance with the 17W75s. For the computer screen shown in *Fig. 4.25*, the values were as follows:

LCR	CR
0.2mH	10.3µF
22.6µF	3Ω
7.5Ω	

Keeping this data for reference, I came up with four different woofer network topographies, each

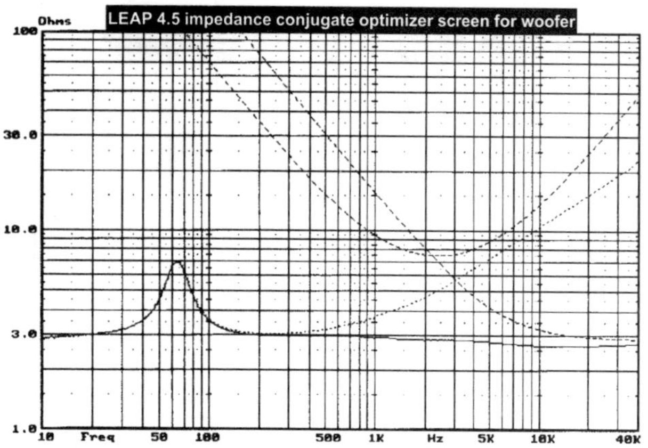

—— Combined Impedance .... Original Impedance - - - Conjugate Nets
DGL Data Info
DGL# 19:17w
imp in box with 100% r19

FIGURE 4.25

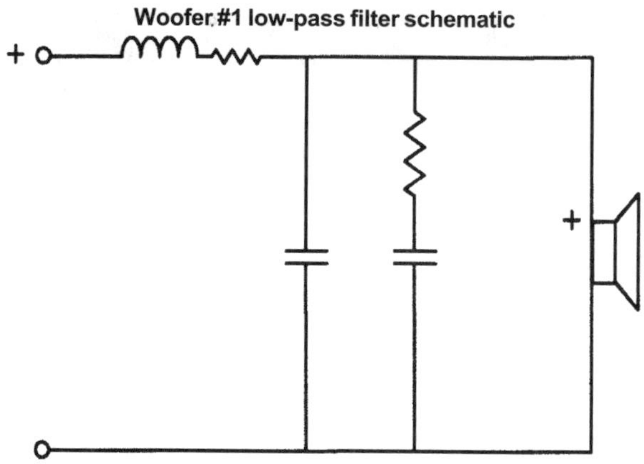

FIGURE 4.26

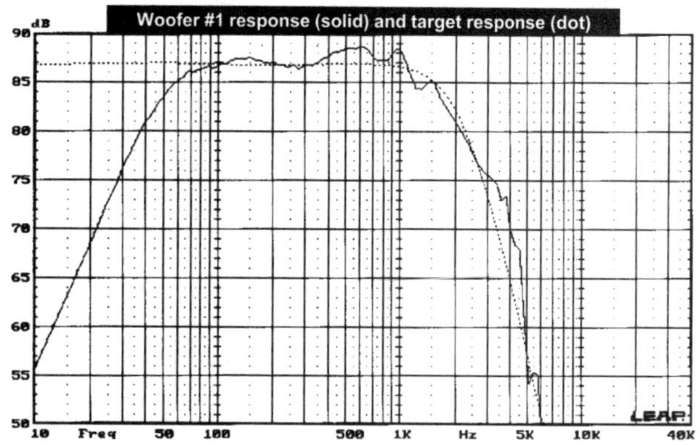

—— Dyn 15W75 1 with one ind. p12
.... Generic lp

FIGURE 4.27

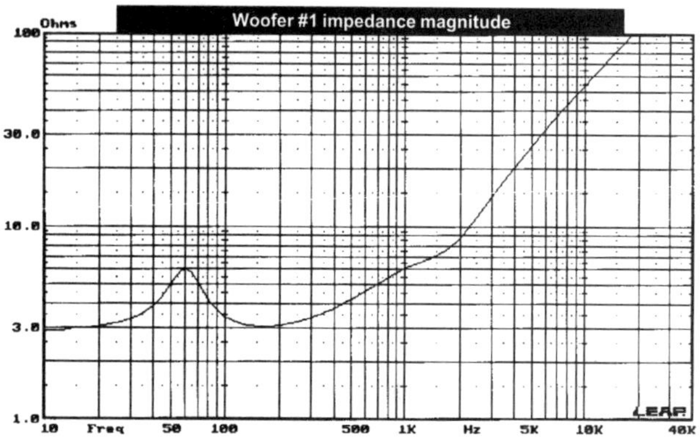

—— Dyn 15W75 1 with one ind. p12

FIGURE 4.28

capable of achieving the appropriate response. These four network sections are:

1. Second-order low-pass + CR conjugate.
2. Modified third-order + CR conjugate.
3. First-order + modified second-order.
4. Third-order + CR conjugate.

I determined component values for each of these filter sections by using a combination of manual "cut and try" on the computer, and running the LEAP crossover optimizer. My chosen target response was 87dB at 2.83V (final measurements were made at 2.02V). Results of this process and the circuit diagrams for each low-pass filter section are depicted in the following figures:

Woofer #	Circuit	Response +Target	Z	Transfer Function
1	4.26	4.27	4.28	4.29
2	4.30	4.31	4.32	4.33
3	4.34	4.35	4.36	4.37
4	4.38	4.39	4.40	4.41

All four circuits are simultaneously compared in *Figs. 4.42-4.44.* Woofer circuit #2 obtained a reasonable fit to the target but had excessive peaking in the transfer function and a low impedance of 2.6Ω. Circuits #1 and #3 both had acceptable impedance and transfer function profiles, but did not match the target function as well as woofer circuit #4, which is the network I chose. *Figures 4.45-4.46* compare the woofer with and without the final network both on-axis and 30° off-axis.

4.60 The Tweeter Crossover

The MDT 33's frequency response curve given in *Fig. 4.18*, shows the SPL about 91-92dB in the 2-3kHz region, which will require about 5-6dB of attenuation to reach the target SPL of 87dB at 2.83V. Even though this tweeter uses magnetic

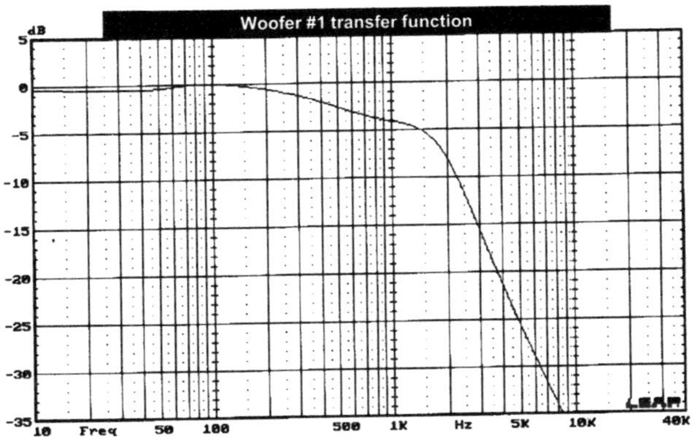

— Dyn 15W75 1 with one ind. p12

FIGURE 4.29

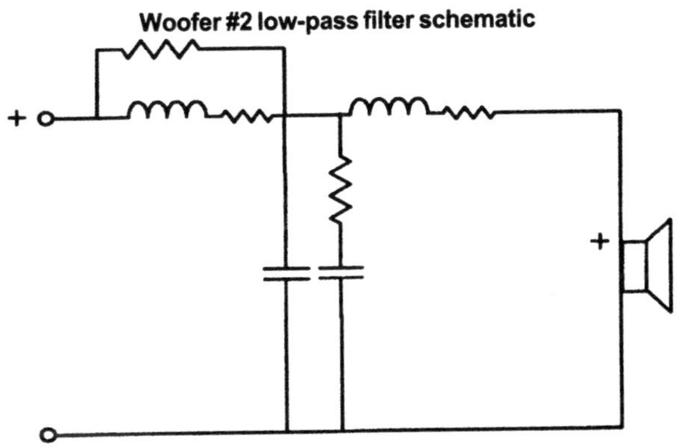

FIGURE 4.30

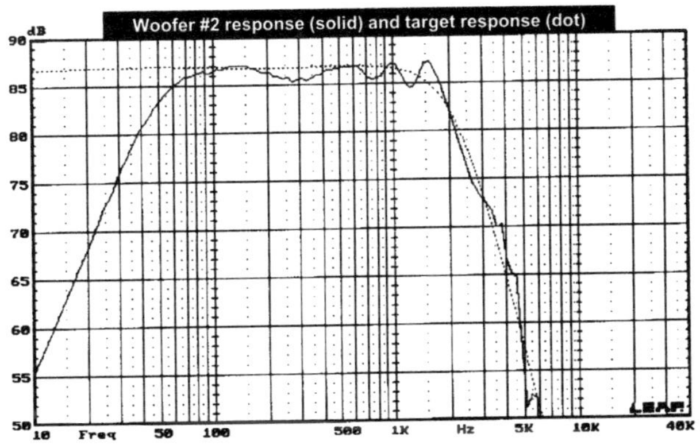

— Dyn 15W75 3 low imp p4
········ Generic lp

FIGURE 4.31

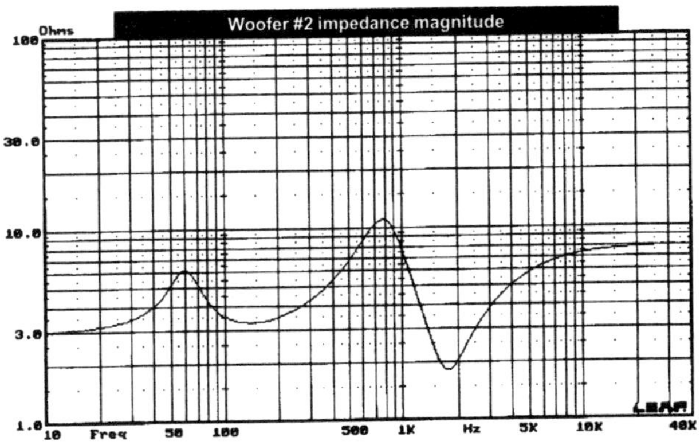

— Dyn 15W75 3 low imp p4

FIGURE 4.32

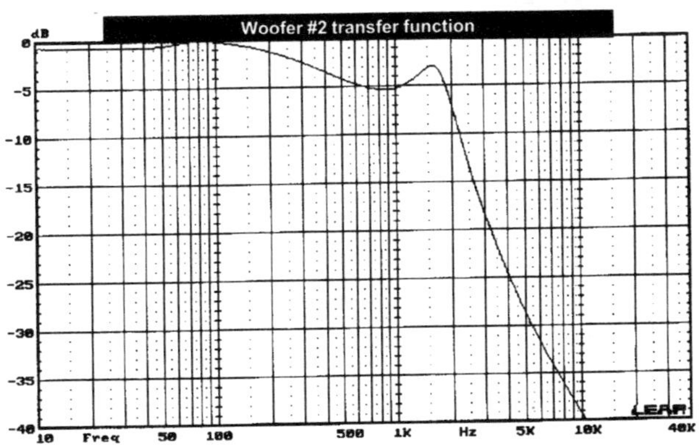

— Dyn 15W75 3 low imp p4

FIGURE 4.33

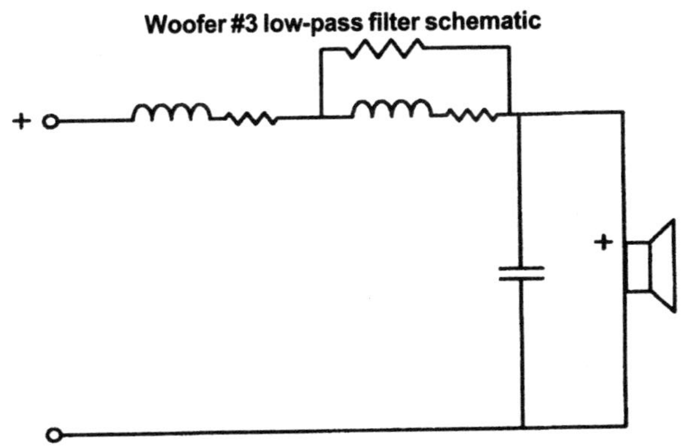

FIGURE 4.34

fluid for resonance damping and voice coil cooling, a substantial impedance peak is still present. I used the LEAP Conjugate Designer to explore incorporating an LCR circuit at the tweeter resonance. The computer screen depicted in *Fig. 4.47* reflects the values below:

L = 0.53mH
C = 79μF
R = 9Ω

Using the SPL target of 87dB at 2.83V, two different high-pass topographies were each optimized to a fourth-order L-R high-pass target response at 2.2kHz. The two high-pass filter sections are as follows:

1. Second-order high-pass + L-pad attenuation.
2. First-order high-pass + LCR and CR conjugates.

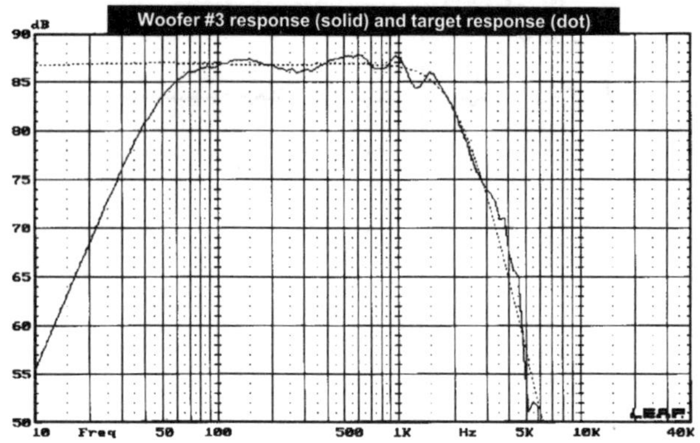

FIGURE 4.35

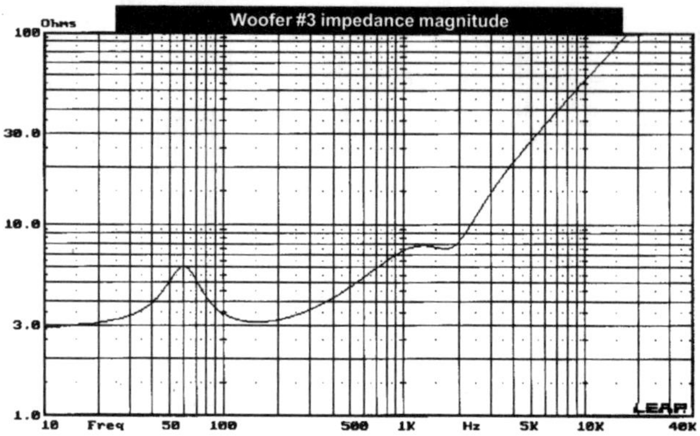

FIGURE 4.36

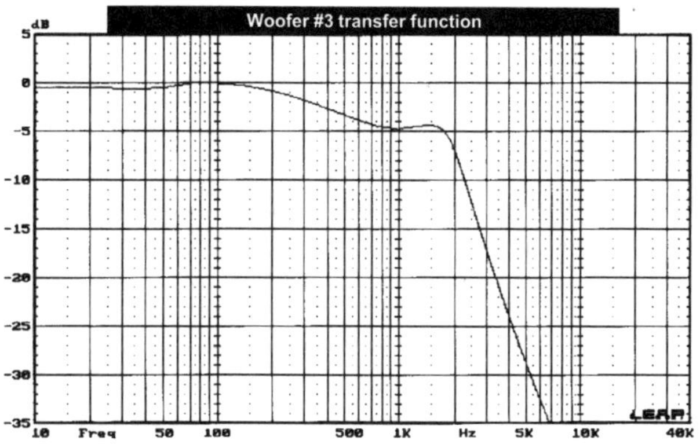

FIGURE 4.37

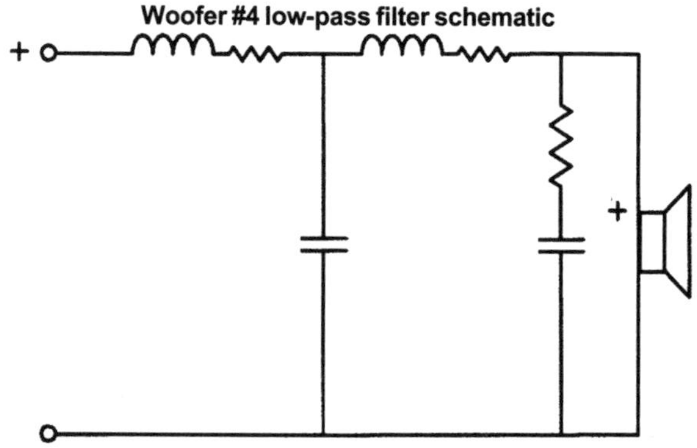

FIGURE 4.38

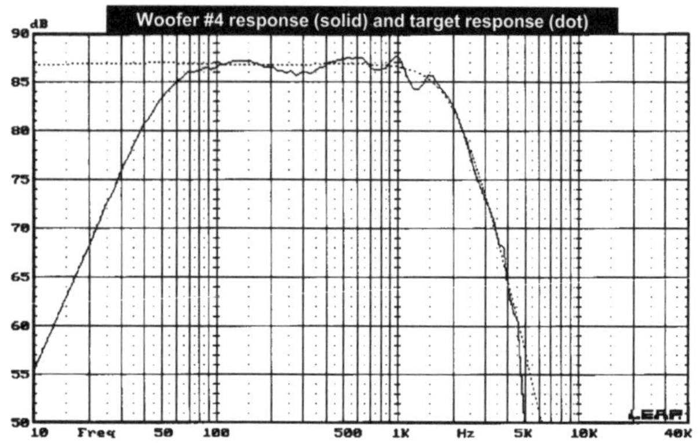

FIGURE 4.39

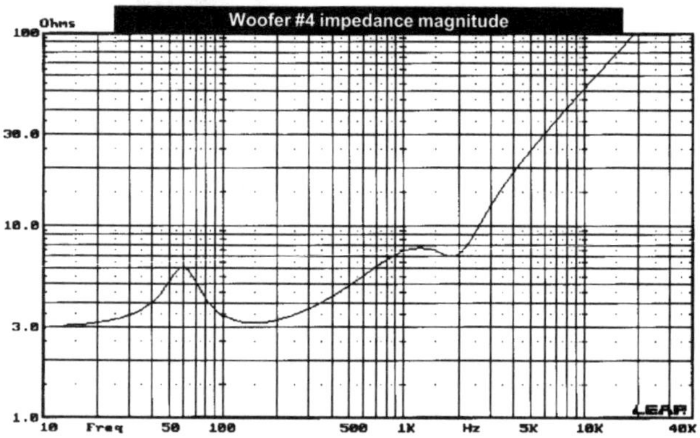

FIGURE 4.40

Both high-pass sections are described in the diagrams listed below:

Tweeter #	Circuit	Response +Target	Z	Transfer Function
1	4.48	4.49	4.50	4.51
2	4.52	4.53	4.54	4.55

Figure 4.53 shows the frequency response of tweeter circuit #2 without the CR network. Switching in the CR network produces the results in *Fig. 4.56*. The before and after comparison is shown in *Fig. 4.57*. When we compare both tweeter circuits, as in *Fig. 4.58*, their response is nearly identical (if we include the CR conjugate in circuit #2). A more complete comparison, using a

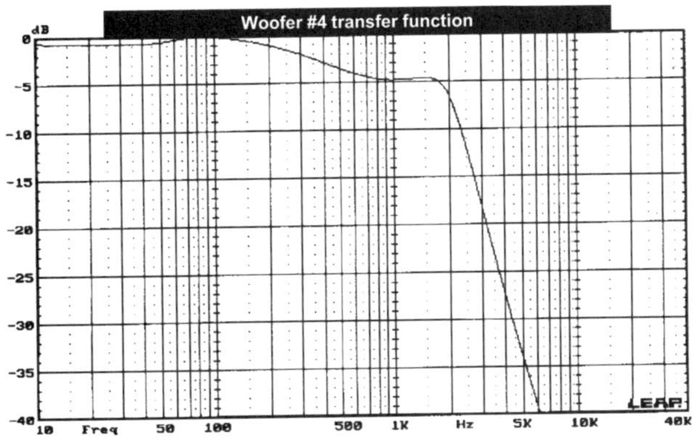

—— Dyn 15W75 1 p2

FIGURE 4.41

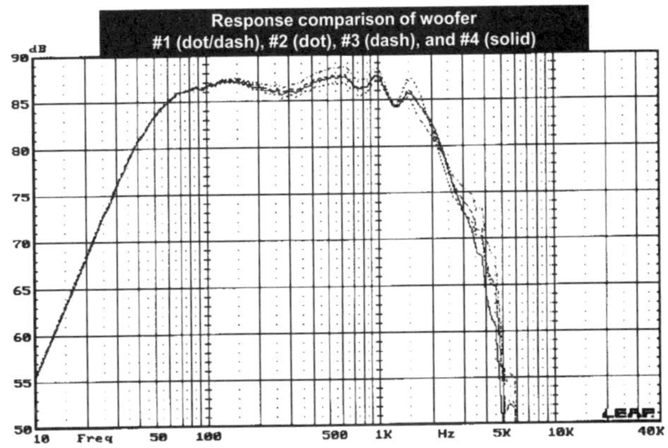

—— Dyn 15W75 1 p2
.... Dyn 15W75 3 low imp p4
- - - Dyn 15W75 2 p3
-·-· Dyn 15W75 1 with one ind. p12

FIGURE 4.42

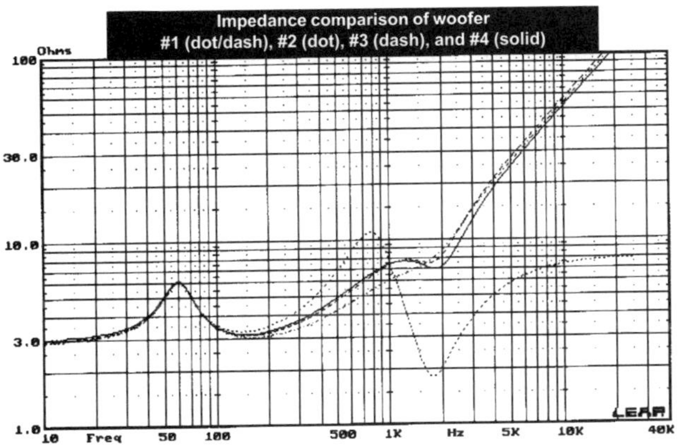

—— Dyn 15W75 1 p2
.... Dyn 15W75 3 low imp p4
- - - Dyn 15W75 2 p3
-·-· Dyn 15W75 1 with one ind. p12

FIGURE 4.43

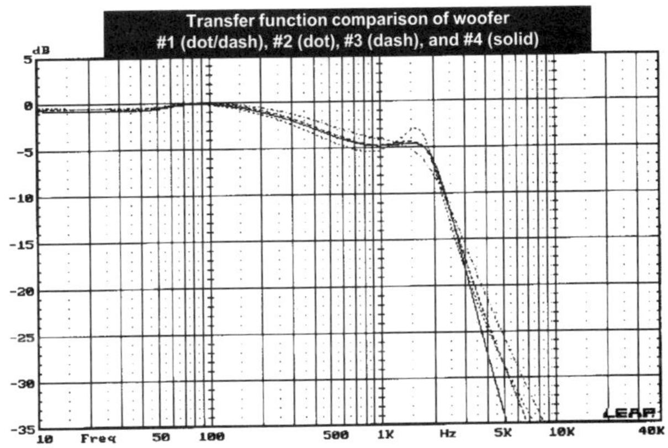

—— Dyn 15W75 1 p2
.... Dyn 15W75 3 low imp p4
- - - Dyn 15W75 2 p3
-·-· Dyn 15W75 1 with one ind. p12

FIGURE 4.44

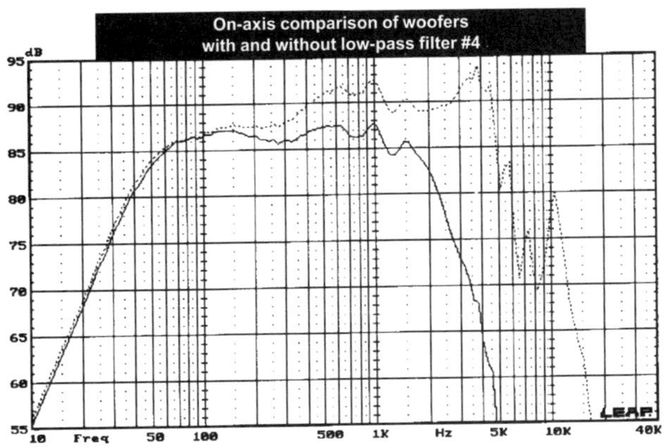

—— Dyn 15W75 1 p2
.... Dyn 15W75 no crossover

FIGURE 4.45

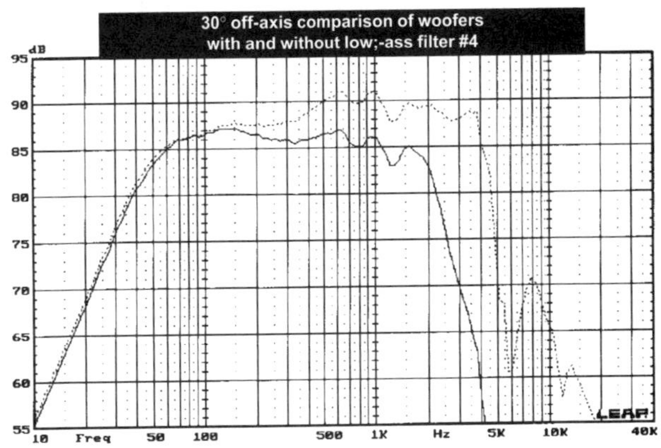

—— Dyn 15W75 1 p2
.... Dyn 15W75 no crossover

FIGURE 4.46

10dB frequency response scale, is shown in *Figs. 4.59-4.61* including response, impedance and transfer function. Obviously circuit #1 has considerably more attenuation in the network stopband in the second octave below the 2.2kHz crossover frequency. Beyond that, the transfer

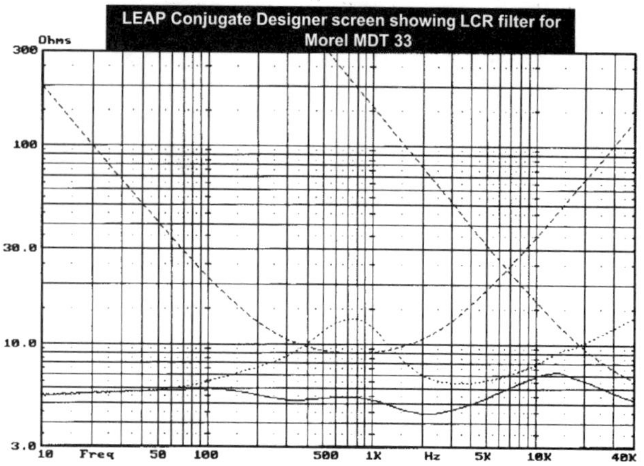

—— Combined Impedance DGL Data Info
········ Original Impedance DGL# 11:MDT33a
- - - Conjugate Nets

FIGURE 4.47

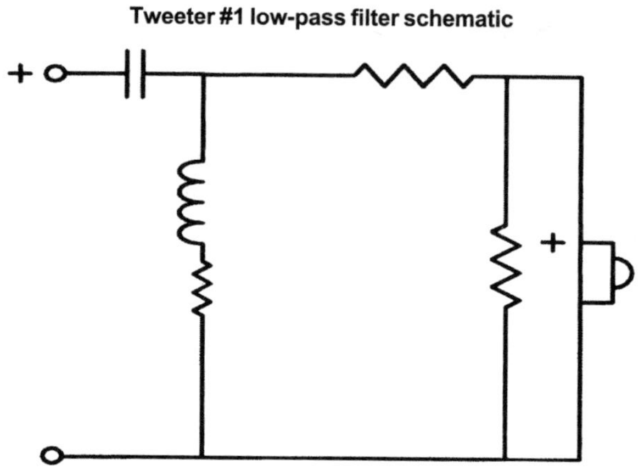

FIGURE 4.48

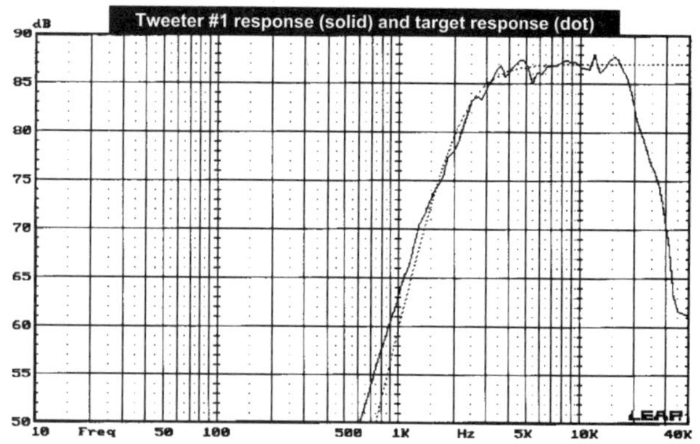

—— MDT33 2 p6
········ Generic lp

FIGURE 4.49

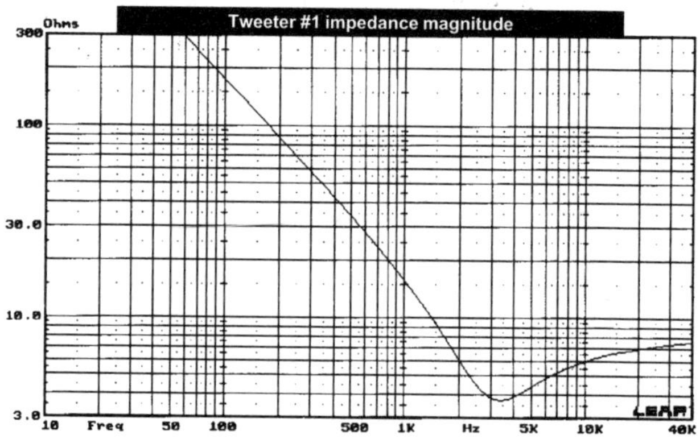

—— MDT33 2 p6

FIGURE 4.50

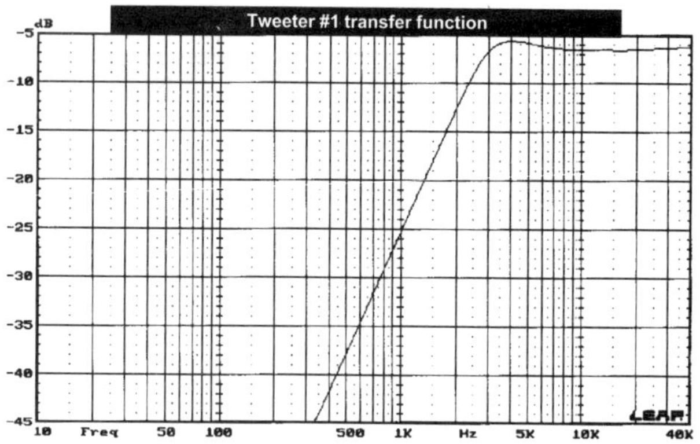

—— MDT33 2 p6

FIGURE 4.51

functions are comparable, with tweeter #1 having a low impedance of 3.8Ω

4.70 System Analysis

The choice of tweeter circuit was determined mostly by the additional attenuation provided in the second octave below crossover by tweeter circuit #1. However, both combinations of woofer circuit #4 and tweeter circuits #1 and #2 are compared in the diagrams listed below:

Tweeter #	In-phase	Out-of-phase
1	4.62	4.63
2	4.64	4.65

Either network pair is usable. Response differences between the two crossovers is fairly trivial, as can be seen in *Fig. 4.66.* The #4 and #1 combination produces less of a null when the tweeter is placed electrically out of phase than combination #4 and #2, but this is not as important a factor, in my opinion, as the increased attenuation provided by tweeter circuit #1. Both circuit combinations were primarily additive except for a small region between 3kHz and 5kHz, which is difficult to correct.

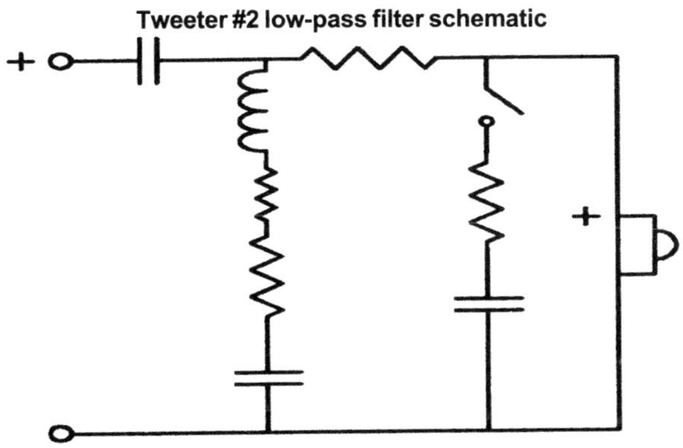

FIGURE 4.52

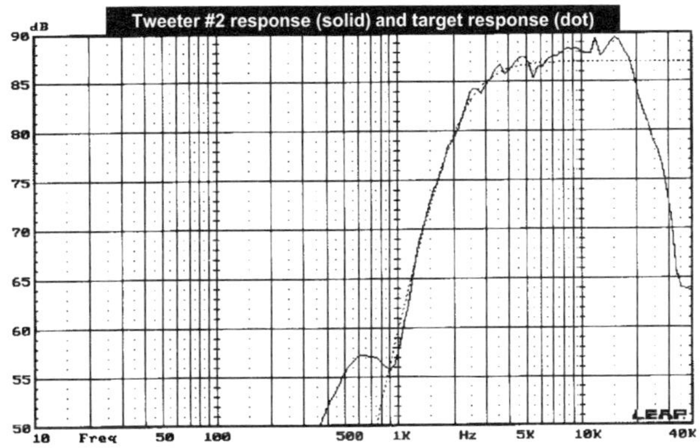

— MDT33 1 with out conj p11
.... Generic lp

FIGURE 4.53

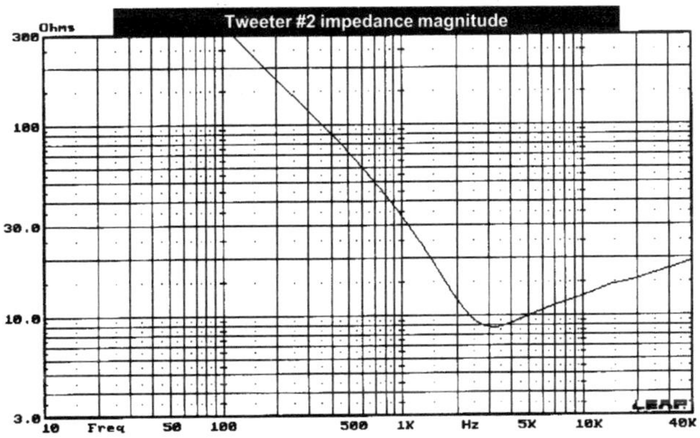

— MDT33 1 with out conj p11

FIGURE 4.54

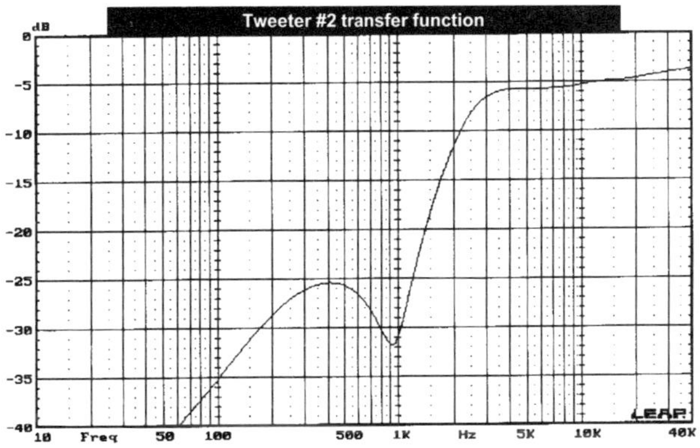

— MDT33 1 with out conj p11

FIGURE 4.55

The final prototype simulation using the woofer #4 low-pass and tweeter #1 high-pass is displayed in *Figs. 4.67-4.70* giving the on-axis response, the 30° off-axis response, the impedance, and transfer function. The tweeter response with and without the network is given in *Fig. 4.71*. The final crossover schematic for this loudspeaker is in *Fig. 4.72*; the component values are listed in *Table 4.1*.

TABLE 4.1
Crossover Component Values

Low-pass section	**Value**
C1	22µF (10µF+12µF) poly
C2	14.9µF (3.9µF+10µF+1µF) poly
L1	0.82mH, 0.19Ω air core
L2	0.3mH, 0.10Ω air core
R1	3.9Ω, 10W
High-pass section	
C3	9µF (5.1µF+3.9µF) poly
L3	0.3mH, 0.10Ω air core
R3	4Ω (two 8Ω 3W metal film)
R4	5Ω (two 10Ω 3W metal film)

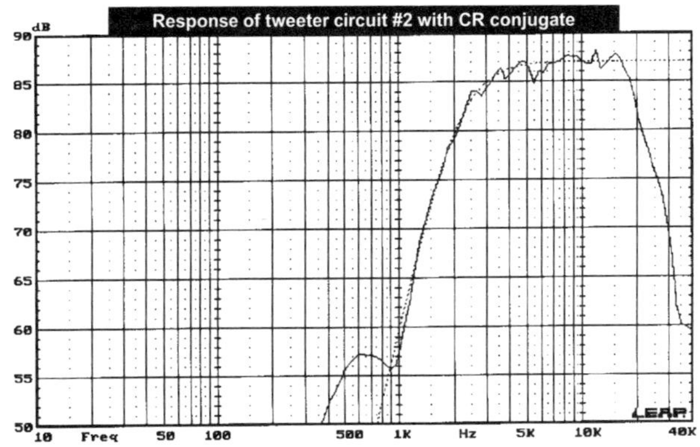

— MDT33 1 p1
.... Generic lp

FIGURE 4.56

Source: All capacitors and inductors: Solen, Inc. Resistors: North Creek Music Systems.

4.80 Measured Performance

Figure 4.73 depicts the final full-range on-axis frequency response of the Dynaudio 17W75 EXT and Morel MDT 33 prototype speaker. This speaker had the most nearly flat response of the

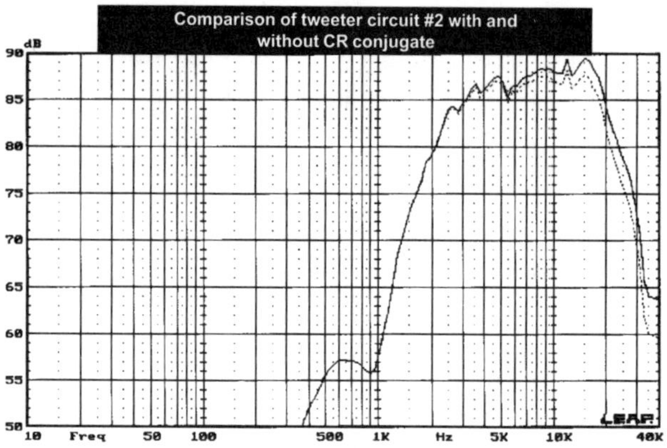

— MDT33 1 with out conj p11
······· MDT33 1 p1

FIGURE 4.57

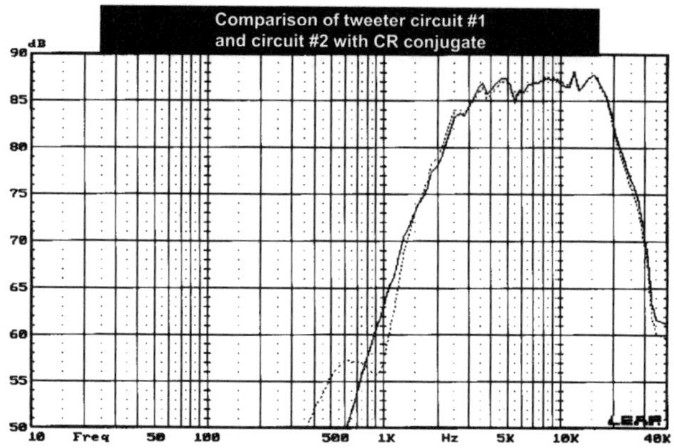

— MDT33 2 p6
······· MDT33 1 p1 w/conj

FIGURE 4.58

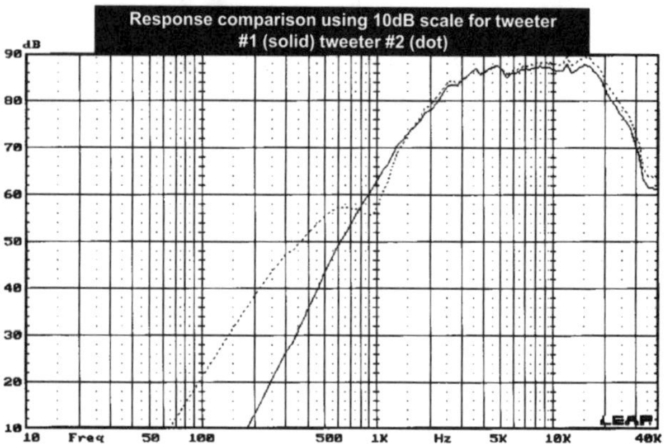

— MDT33 2 p6
······· MDT33 1 p1

FIGURE 4.59

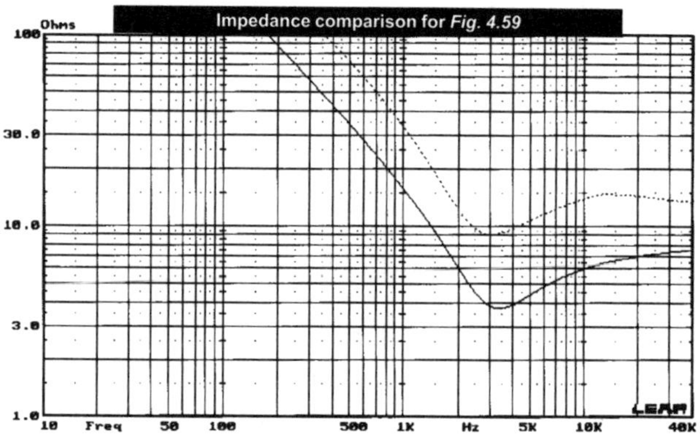

— MDT33 2 p6
······· MDT33 1 p1

FIGURE 4.60

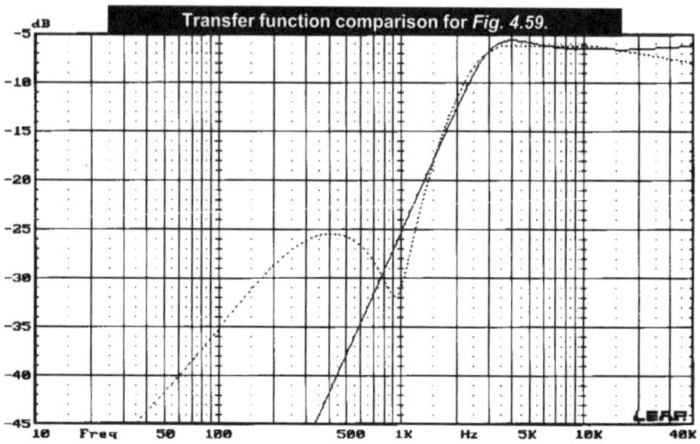

— MDT33 2 p6
······· MDT33 1 p1

FIGURE 4.61

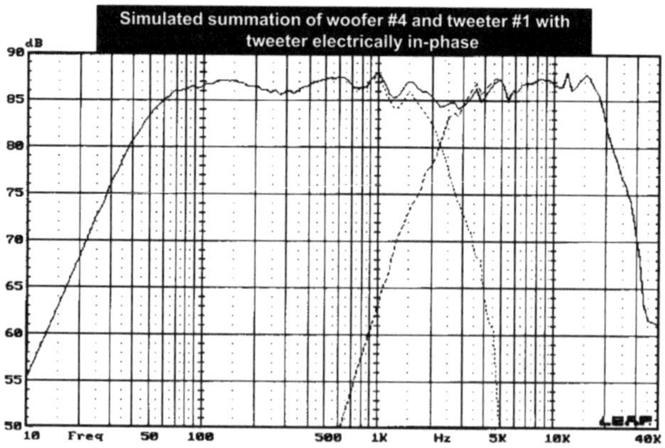

— 4+7
······· Dyn 15W75 1 p2
- - - MDT33 2 p6

FIGURE 4.62

book's four designs and was ±1.73dB from 55Hz to 20kHz, and ±3dB from 45Hz-21.6kHz. I made all final measurements at 2.02V, which would be 1W at the 4Ω load. The sensitivity of this speaker, even with a double woofer section, is 84dB 1W/1m. The horizontal off-axis response at 0°, 15°, and 30° in *Fig. 4.74* shows the speaker to be within ±2dB out to 15kHz at 30° off-axis.

The response out to 45° in *Fig. 4.75* droops somewhat, but still provides a very flat profile

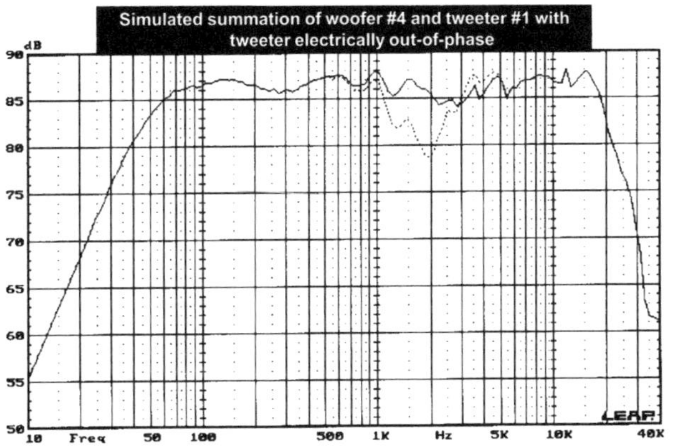

—— 4+7
······· 4+7 out of phase

FIGURE 4.63

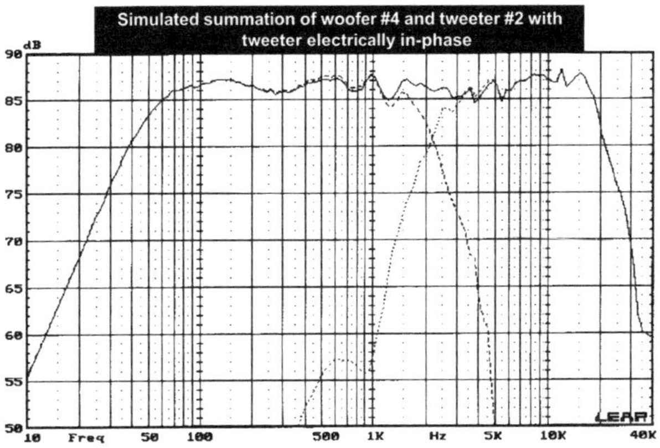

—— 3+4
······· MDT33 1 p1
- - - Dyn 15W75 1 p2

FIGURE 4.64

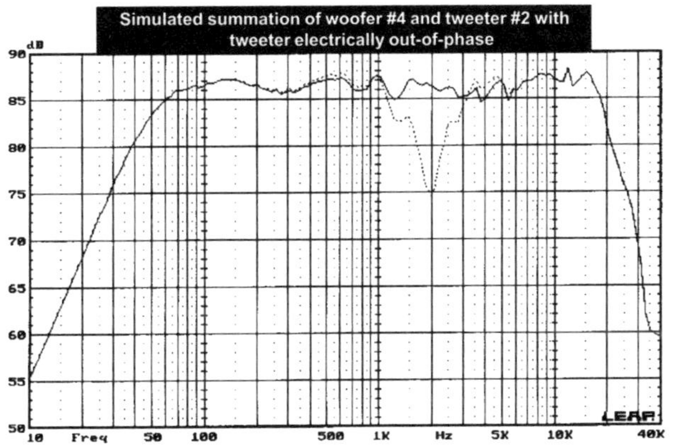

—— 3+4
······· 3+4 out of phase

FIGURE 4.65

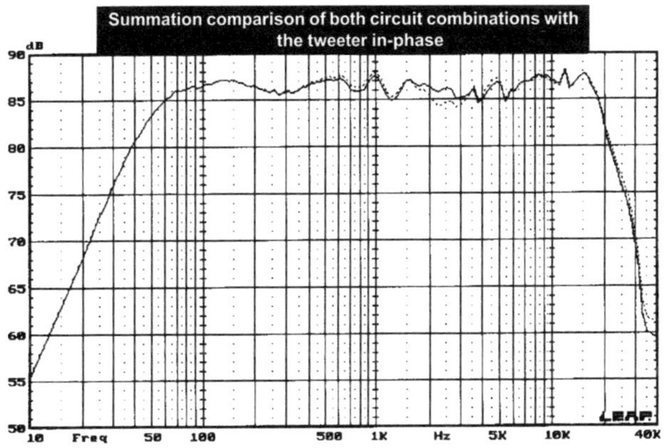

—— 3+4
······· 4+7

FIGURE 4.66

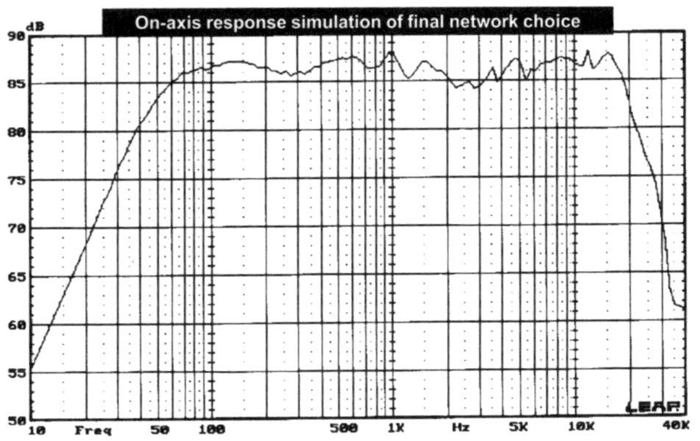

—— Dyn 15W75 1 p2
······· MDT33 2 p6

FIGURE 4.67

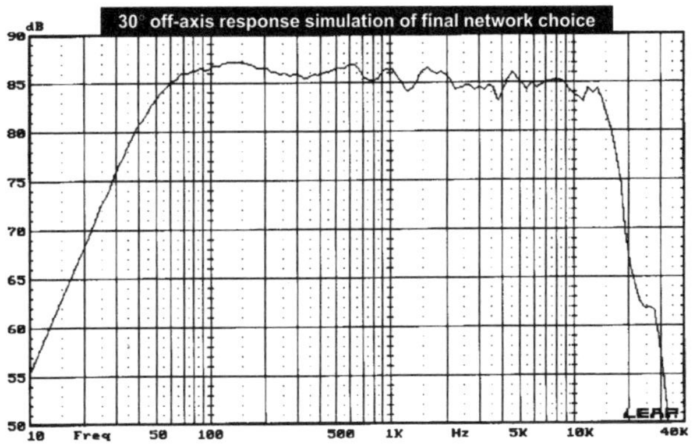

—— Dyn 15W75 1 p2
······· MDT33 2 p6

FIGURE 4.68

except for the decline in upper frequency response above 8kHz. The midband frequency SPL is maintained quite well out to 90°, evident in *Fig. 4.76*, demonstrating the excellent dispersion characteristics of the Dynaudio 17W75. The overall horizontal off-axis response of this speaker is quite good.

The vertical off-axis response taken up and over the top of the speaker at 0°, 15°, and 30° is illustrated in *Fig. 4.77.* Notice that the dip in the

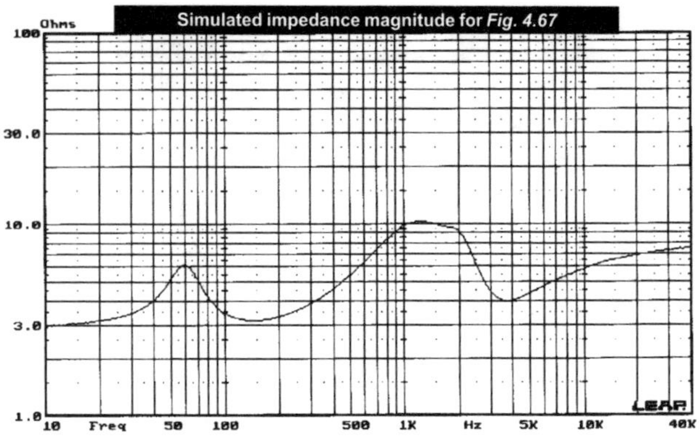

— Dyn 15W75 1 p2
········ MDT33 2 p6

FIGURE 4.69

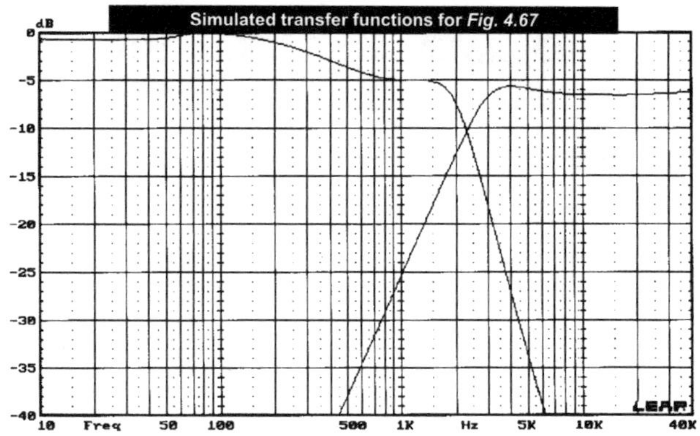

— Dyn 15W75 1 p2
········ MDT33 2 p6

FIGURE 4.70

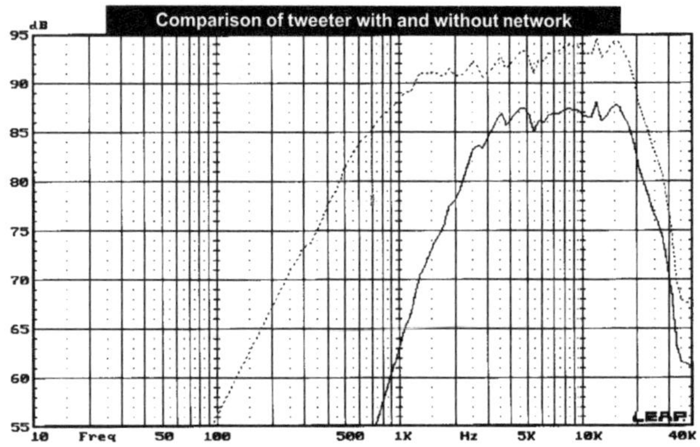

— MDT33 2 p6
········ MDT33 no crossover

FIGURE 4.71

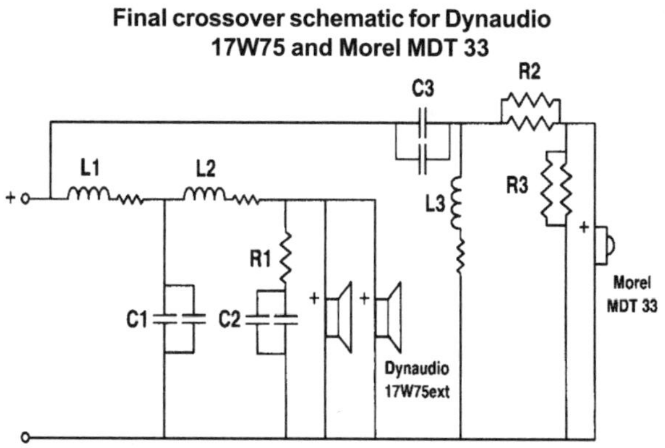

FIGURE 4.72

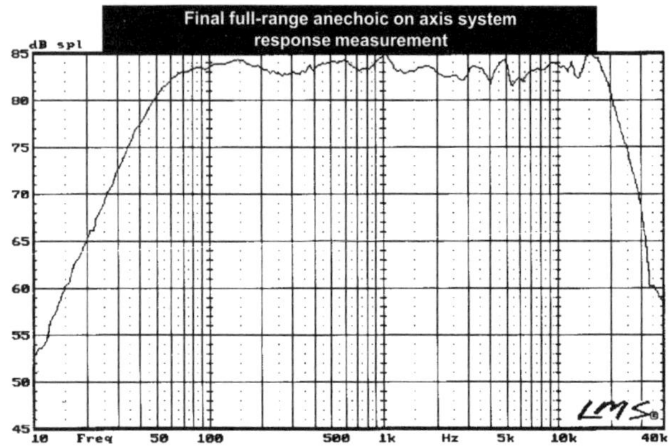

— Full range

FIGURE 4.73

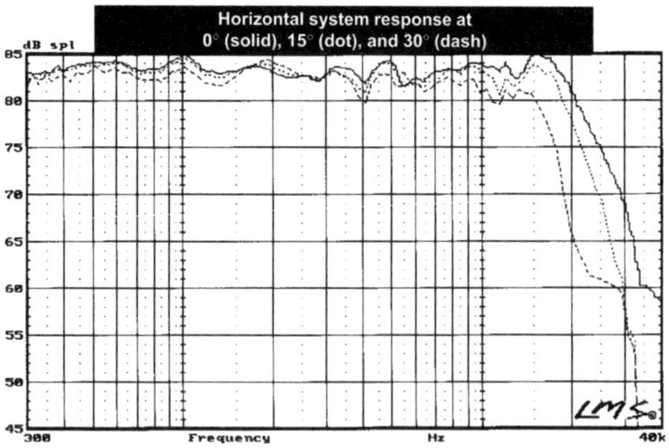

— Full range
········ M3a 15
- - - M3a 30

FIGURE 4.74

response at 30° off-axis is lower in frequency than the dip at 15°, certainly more so than was observed in the other two driver two-way speakers. This is due to the interference effects caused by having two woofers radiating in the low frequency range that are separated by the tweeter. This same effect is apparent (*Fig. 4.78*) in the vertical off-axis response taken below the speaker.

Left and right channels were well-matched to within 0.5dB except for one small area at 1.5kHz

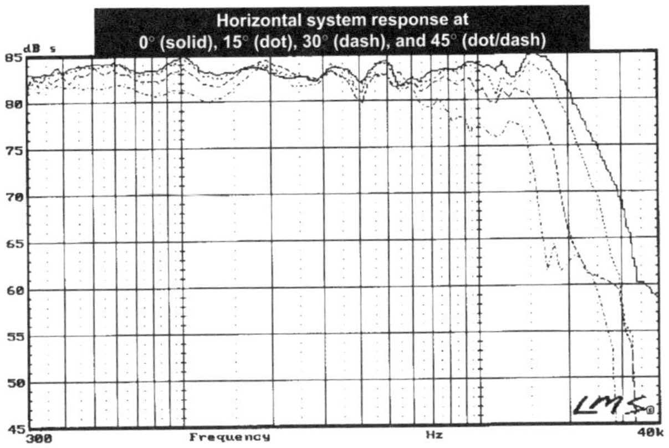

— Full range
········ M3a 15
- - - M3a 30
-·-· M3a 45

FIGURE 4.75

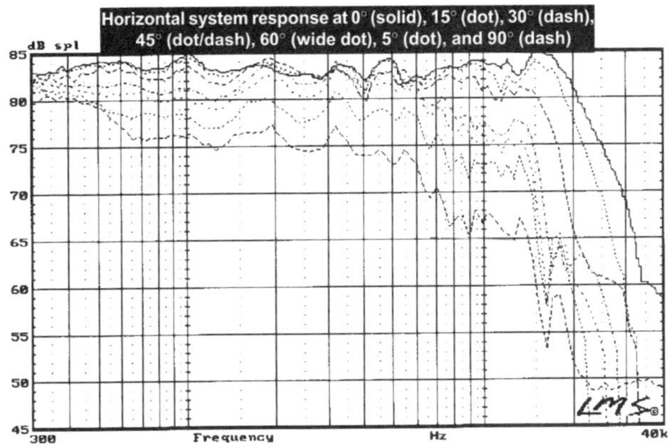

— Full range
········ M3a 15
- - - M3a 30
-·-· M3a 45
· · · M3a 60
········ M3a 75
- - - M3a 90

FIGURE 4.76

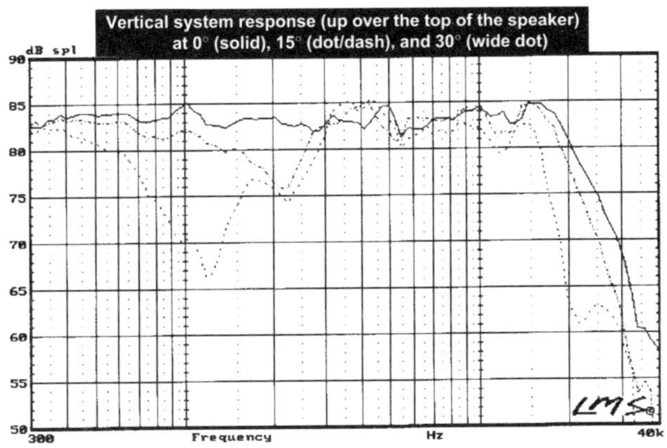

— M3a 0h
-·-· M3a +15
· · · M3a +30

FIGURE 4.77

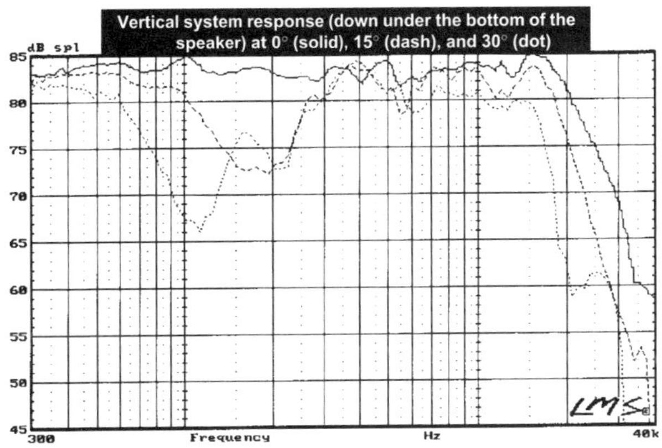

— Full range
- - - M3a -15
········ M3a -30

FIGURE 4.78

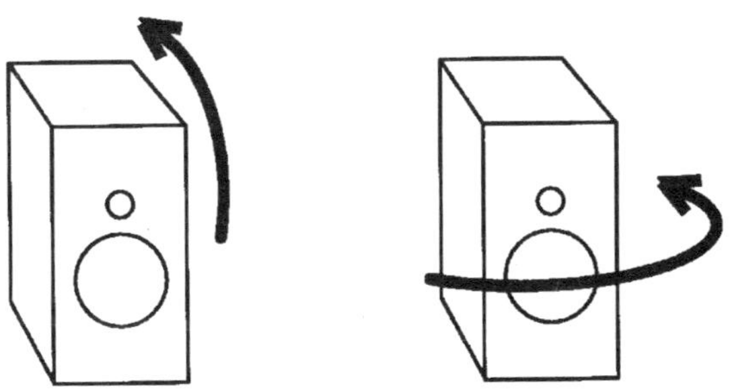

FIGURE 4.78B

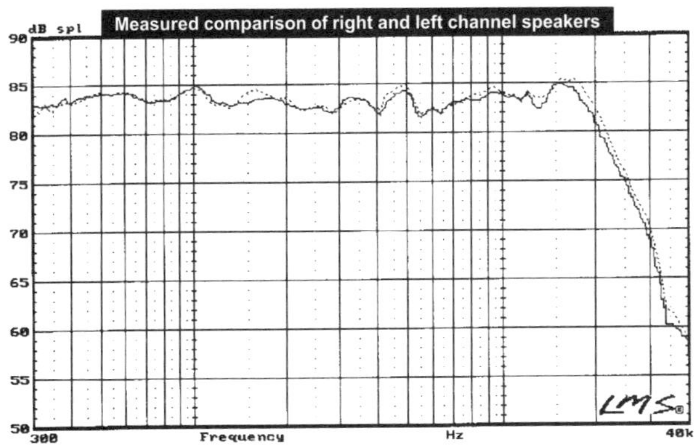

— Full range
········ M3b 0

FIGURE 4.79

as depicted in *Fig. 4.79*. This close matching should be easy given the QC done on the matched Morel tweeters and the normal QC level in Dynaudio components.

Figure 4.80 shows the comparison of the in-phase system response, and the system response with the tweeter electrically out-of-phase, plus the individual woofer and tweeter responses. The out-

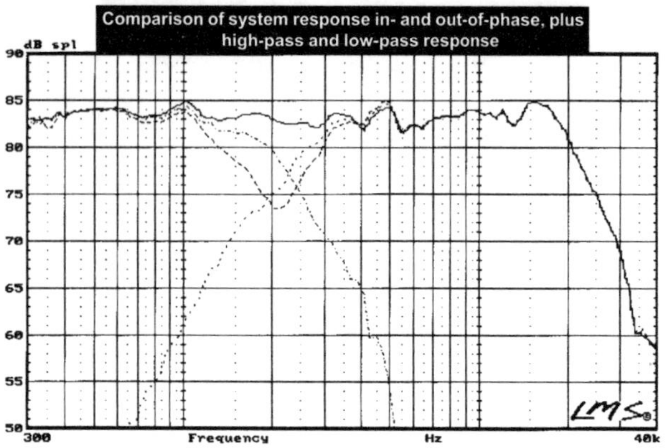

— Full range
-·-· M3a wof
· · · M3a tw
- - - M3a rev

FIGURE 4.80

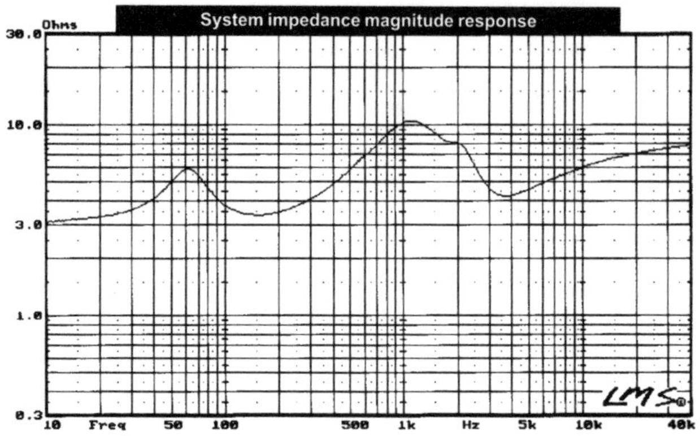

— M4 sys imp

FIGURE 4.81

Complex impedance curve (Nyquist plot)

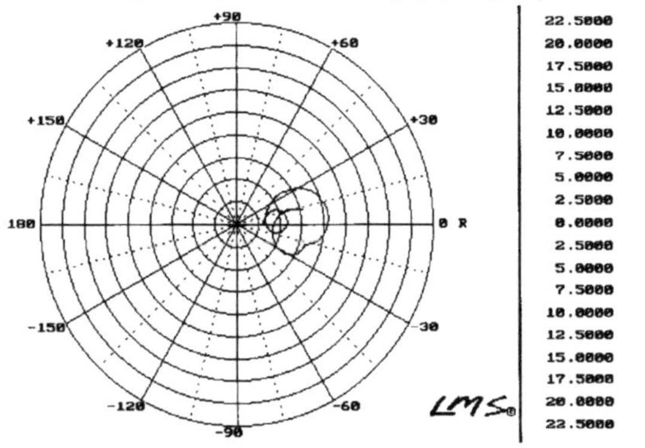

- - - M4 sys imp

FIGURE 4.82

MLSSA cumulative spectral decay plot

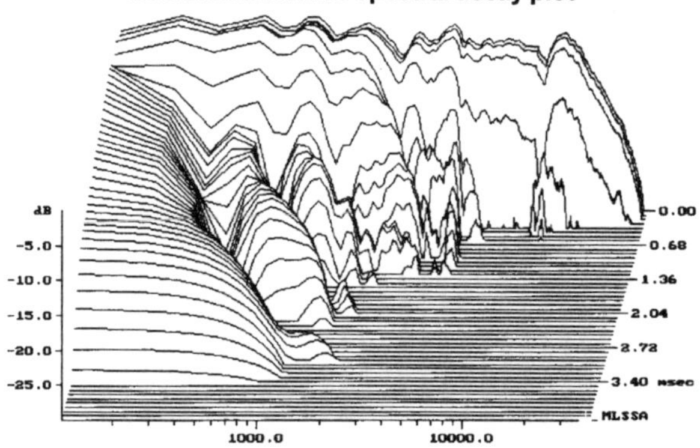

FIGURE 4.83

of-phase null occurs only slightly lower than the 2.2kHz crossover frequency, is fairly symmetrical, and almost all the crossover region is additive except for the 5-6kHz range. This is close to the LEAP predictions, with deviations probably due in part to the inexact mike locations. Because of the dual woofer setup, mike position is critical and small changes can cause larger differences than when a single driver is in each frequency range.

Figures 4.81 and *4.82* display the impedance and Nyquist complex impedance curves. The low impedance point in *Fig. 4.81* is at 125Hz at 3.4Ω, and close to 3Ω at 10Hz, however today's new high current amplifiers should have no trouble with this low impedance. The maximum inductive phase angle is 34.4° at 520Hz and the maximum capacitive phase angle is 30° at 2,482Hz, neither of which would be difficult loads for any amplifier. This speaker, however, should not be paralleled with another loudspeaker because of its low impedance at lower frequencies.

The last performance graph is the MLSSA cumulative spectral decay curve depicted in *Fig. 4.83*. No sharp resonances are evident and the speaker is well-damped at frequencies above 2kHz attenuated to -30dB by 2ms.

4.90 Adding a Subwoofer

Because of the limited dynamic range of the system's woofers, I strongly recommend using a subwoofer similar to the one described in *Chapter 3*. However, the performance without the subwoofer is really quite good so long as you don't push it to 100-105dB SPL levels with the spectacular low-frequency transient material from some CDs (such as the cannon shots on Telarc's 1812 Overture). However, if you go to the trouble of producing a speaker with this kind of detail, you might as well go all the way and give it a really deep, low-distortion bottom octave by biamping to a subwoofer.

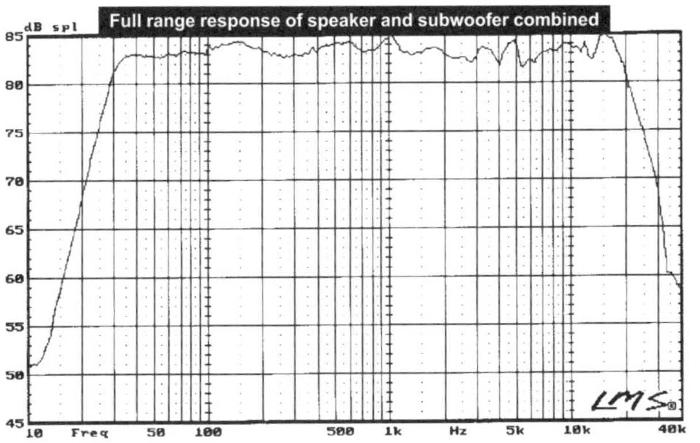

— M4 + sub

FIGURE 4.84

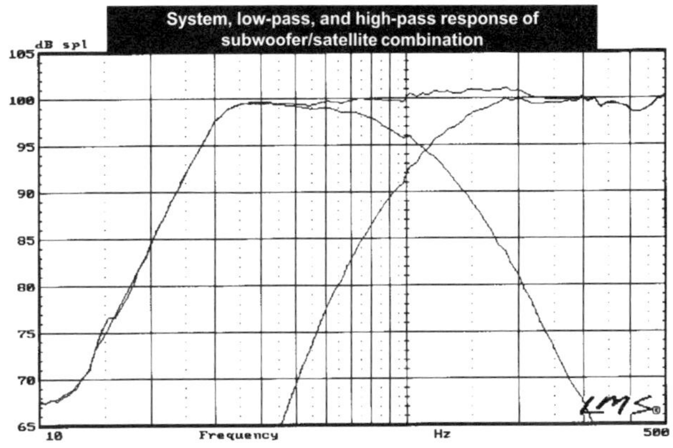

— M3 2 sub
······· M3 2 sat

FIGURE 4.85

MLSSA accelerometer CSD plot of box side panel

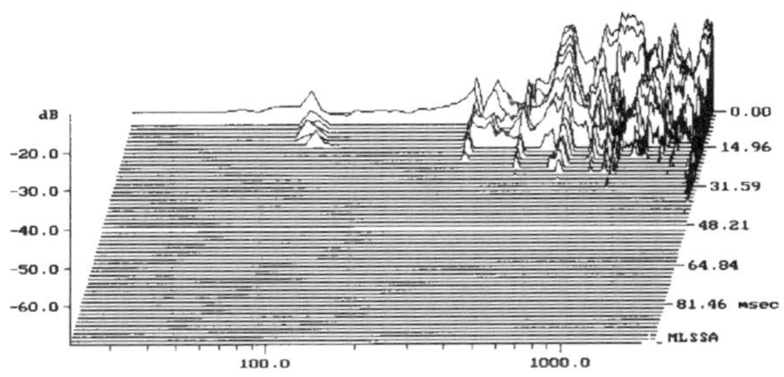

FIGURE 4.86

MLSSA accelerometer CSD plot of box top panel

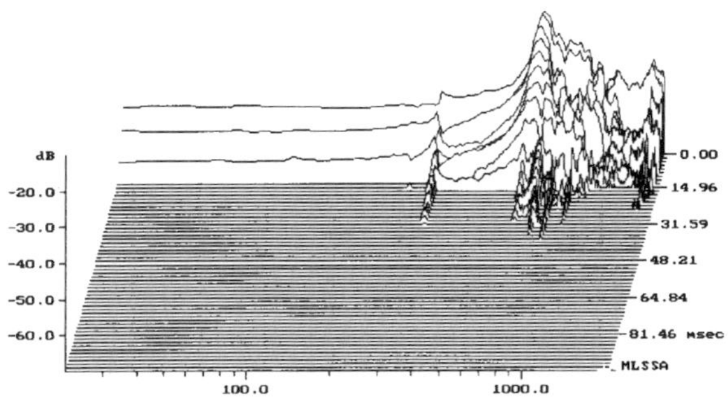

FIGURE 4.87

Figure 4.84 illustrates the full-range frequency response of the prototype using the TDM 24CX-2 electronic crossover and the Vifa M22WR subwoofer. The high-pass and low-pass response is compared in *Fig. 4.85*. The slight rise at 140Hz could be tuned out using the crossover controls, and are generally "tweaked" depending on the individual room acoustics.

4.100 Cabinet Performance

Construction of the Dynaudio/Morel speaker is similar to the first two in *Chapters 1* and *2*, except that this design does not use a Sonotube baffle. The box is constructed with 1" MDF which has a 1.75" front baffle, a 0.75" internal MDF baffle, extensive dowel cross bracing, and extensional damping material. This cabinet is quite substantial and weighed 75 lbs.

An accelerometer was used to test the effectiveness of the cabinet design using both the MLSSA and LMS analyzers. Results using MLSSA are shown in *Figs. 4.86-4.88* for the side, top and rear cabinet areas. *Figure 4.89* shows the LMS measurement of the enclosure side and is divided by frequency twice (compared with the production speaker described in *Chapter 1*), which results in converting the measurement from acceleration to excursion. These plots can be readily compared to the other data provided in previous chapters, however, if you use the techniques described throughout this book you will have a well-damped cabinet.

4.110 Construction Details

The cabinet construction details are diagramed in *Fig. 4.90*. They are nearly identical to the construction details described in *Chapters 1* and *2*. Although the damping characteristics of the cabinet are excellent, the construction procedures are simple. The top, bottom, rear baffle and left and right walls are 1" MDF, and assembled using carpenter's glue and staples (nails). The four walls are simply butt jointed for ease of assembly.

The angled internal baffle separates the two woofer cavities and limits internal reflections and standing wave modes. The 0.75-inch-thick plate was installed as shown in *Fig. 4.90*.

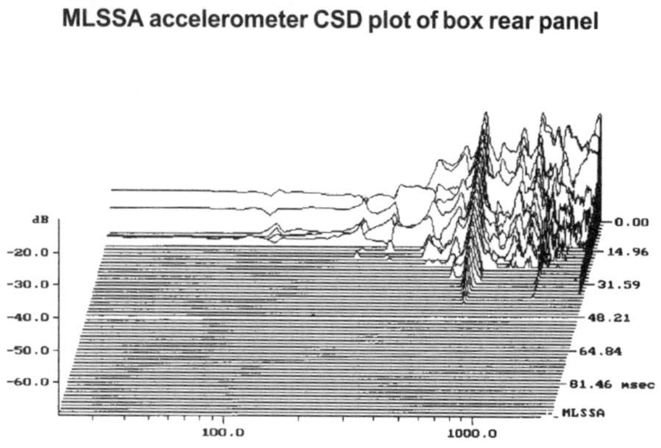

FIGURE 4.88

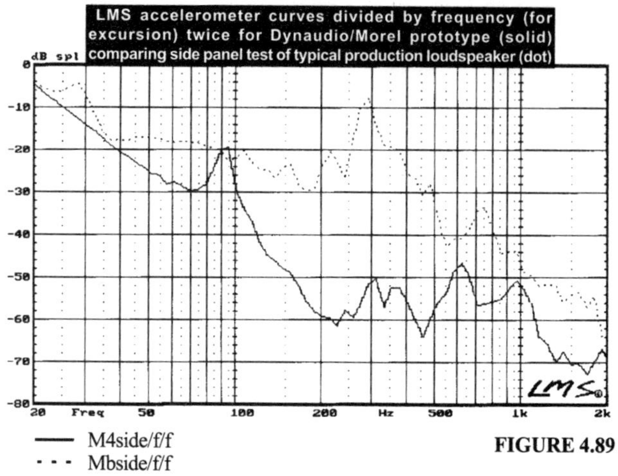

FIGURE 4.89

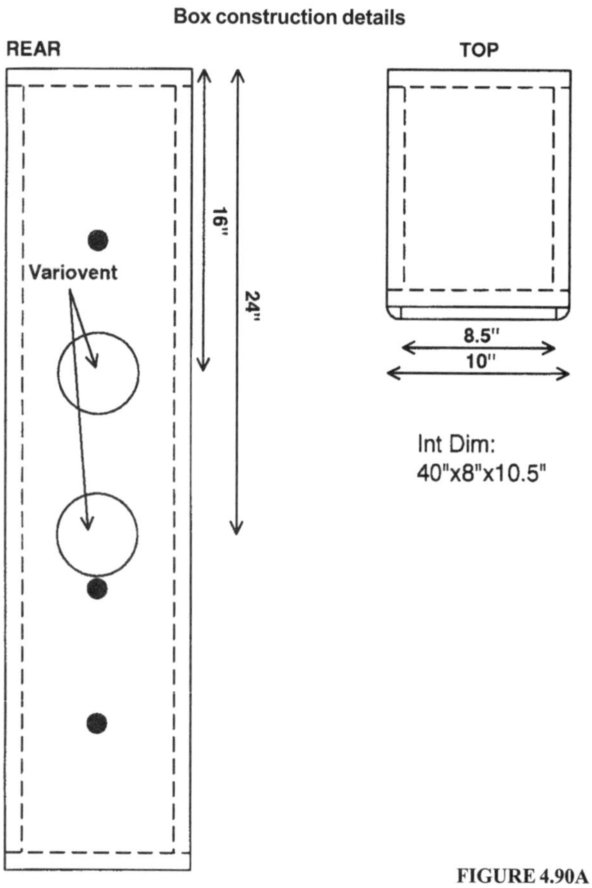

FIGURE 4.90A

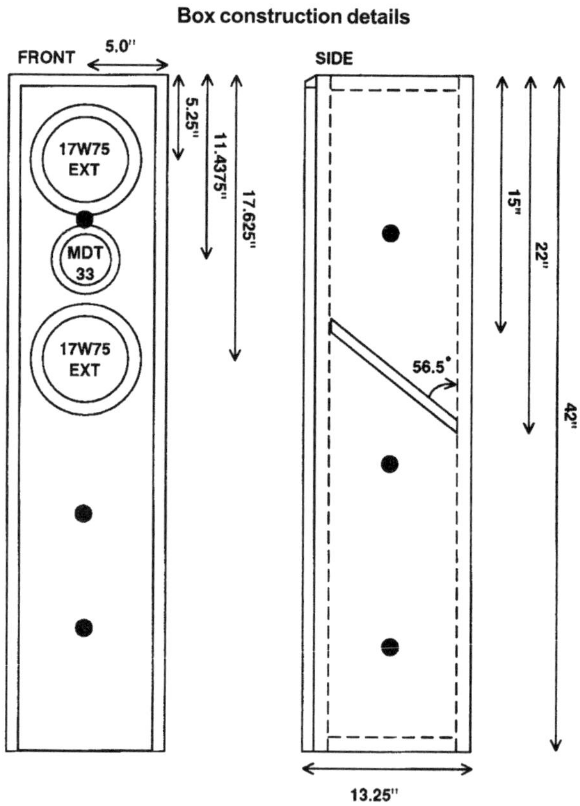

FIGURE 4.90B

The six 1.125" hardwood dowel cross braces are located as shown by the small black circles in the box construction diagram. Cutting the dowel to exact length can be a problem depending on your woodworking skills. I drilled a shallow 0.25" deep countersink into each wall to position the dowels, and cut them 1/4" longer than the enclosure dimension. Apply construction adhesive to the dowel insets, and set the dowels in position during box assembly. When the adhesive sets, the braces are securely tied to the enclosure.

This avoids the too long dowels bowing the enclosure outward, or too short ones pulling the walls inward. Although dowel cross bracing is effective, it's easier than using shelf-type bracing. If you are a confident woodworker, you can choose this type of bracing as long as you calculate the small added volume and increase the box height dimension accordingly.

The front baffle is 1.75" thick and was made from an inner 1" MDF layer and an outer one of 0.75" MDF. Assemble the baffle by first gluing and stapling the 1" layer to the box, and then the 0.75" layer. I routed the 0.75" outer section to countersink the tweeter flush with the baffle surface (the Dynaudio woofers need not be inset). To prevent acoustical loading of the woofer by the 1.75-inch-thick baffle I routed the inner layer with a curved 45° bit on the inside of the woofer hole. The 0.75" baffle layer was inset 0.75" on both sides and top to accommodate the 0.75" oak quarter-round trim. The detail of the oak top trim piece, which extends across the top of the 0.75" layer of the front baffle, is illustrated in *Fig. 2.97* in *Chapter 2*. Cut the oak quarter-round at the same angle as the oak top trim piece to produce a smooth angle.

Once the cabinet was assembled, I painted the baffle front and enclosure bottom. I cemented oak veneer to the sides, the lower half of the baffle front, top and rear, and glued the oak trim in place. Internal wall panels (sides, front baffle, top and bottom) were further damped with two layers Antiphon Type A-13 extensional material, a 1/16-inch-thick black bituminous felt and clay material with self-adhesive backing. Approximately 50-60% of each wall was covered. An alternative, which is as effective if not more so, is the Black Hole Pad, which is thicker material and a single layer is sufficient.

To damp enclosure standing waves, I lined all inner surfaces with Owens Corning Type 703 Acoustic Fiberglass, except the Variovents. I filled the remaining space with R19 pink household fiberglass, except behind the woofers opposite the Variovents. Use a face mask and gloves when you do this particularly nasty chore. Finally I filled the open area between the woofers and the Variovents with teased Acousta-Stuf (>0.5 lbs/ ft^2).

Well-Nut inserts again were the mounting hardware for the woofers and tweeter. Their rubber inserts often serve to damp electric motor vibrations. They mount simply and allow easy driver removal and also help isolate the drivers from the cabinet. I added a foam tape gasket as well, the same type used for weather stripping. North Creek also offers a better neoprene rubber gasket tape.

Unlike most speakers the crossovers were not mounted inside the enclosure, but in the vertical stem of the stand. This not only isolates the components from vibration and driver magnetic fields, but also allows for a much larger component layout, allowing wider placement of the inductors.

Front Baffle

Two terminal sets (one set each for the woofer and tweeter) are required on the enclosure rear.

I used a small ACI round terminal cup which fits a 2 7/8" hole, with two gold binding posts. I added two more binding posts identical to ACI's. The four provide direct connection to both the woofers and the tweeter. Since this is a sealed box, seal all internal joints with a small bead of silicone to ensure an airtight enclosure. Also seal the wiring lead hole between the woofer cavities.

I made the grille frames from 0.25" MDF with 0.25" thick standoffs for use with snap type plastic grille fasteners. The standoffs were half-inch-square pieces of 0.25" MDF glued to the frame. The standoffs add an attractive look, as well as the required thickness to mount the plastic fastener. Black double-knit cloth stretched over the frame was fastened with 3M spray fabric adhesive. This format works well, but you are free to alter this part of the design.

Construction detail for the stands is shown in *Fig. 4.91.* I used 0.75" MDF for both the base and the vertical stem, finishing in black laminate with oak iron-on tape on the edges. Brass cone-type feet were added to the base's bottom by drilling shallow countersink holes. The base was attached

Construction detail of speaker stand

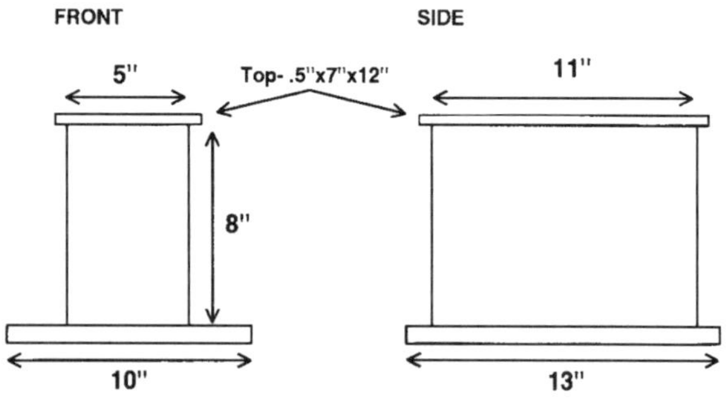

FIGURE 4.91

Photograph of crossover layout

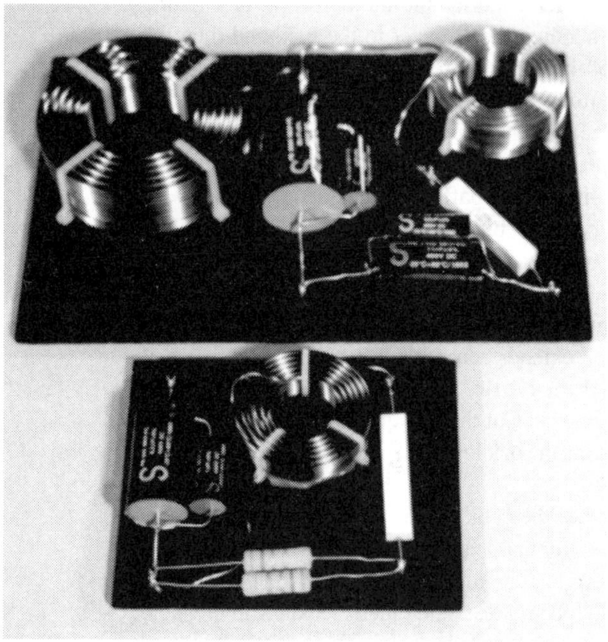

FIGURE 4.92

to the vertical stem using 1.5" black wood screws. Like the speakers, the vertical stem was also finished with black laminate. I lined the stem's inside walls with Black Hole Pad to reduce vibration transferred from the speaker cabinet.

I chose a rectangular blank terminal cup to mount three pairs of gold binding posts, for the input, the high-pass section for the tweeter, and for the low-pass section for woofer respectively. I added the tweeter high frequency contour switch to the cup also. Obviously, you may add another pair of terminals if you opt for biwiring. Banana plug jumpers connect the crossover in the base to the drivers in the speaker enclosure.

Although I have not shown separate detail of the stand's top piece, I added small rubber pads, the kind used as feet for electronic equipment to help isolate the speaker from the stand. The top piece is fastened with wood screws to the stand stem. I added rope caulk, the kind used to seal windows and doors, as a gasket between the top and the vertical stem, for additional isolation.

The crossover is mounted on 0.25" hardboard cut to fit the side wall of the inside of the stand stem. High- and low-pass sections are on separate boards and cannot be placed on a single board as were the crossovers of the other three speakers. Mount the woofer section on one side of the stand wall with the inductors toward the bottom. Mount the tweeter section at the top of the stand's opppsite side wall. Because I mounted the inductors this distance apart, I did not place adjacent inductors at right angles and perpendicular to each other. Even though mounted in the same plane, I measured no inductive coupling between individual sections.

However, you can get another 10-20dB attenuation by mounting them at 90° angles, but use some kind of strapping to hold the inductors firmly in place. Component layout is not critical except for inductor placement, as shown in *Fig. 4.92*. The woofer section was mounted to the side wall using silicone adhesive, as was the tweeter section. All wiring in both stand and cabinet was done with Monster cable for the woofer and Monster XP for the tweeter.

I wired crossover components point-to-point using brass split rivets as connection points for soldering, which also provides a mechanical board mounting for capacitors and resistors. To locate the split rivets lay out the components on the board, and mark the junction points with a pencil. Next drill the board and insert the rivets. Inductors were mounted on the crossover boards with silicone adhesive.

GLOSSARY OF TERMS

Accelerometer. An accelerometer is a transducer used to measure vibration intensity. Its application extends to virtually anything which vibrates, such as a jet engine's housing or a duct in an HVAC system. Accelerometers are mostly used to reveal the secondary vibration of a loudspeaker cabinet when it is set into motion by a woofer's action. They are also sometimes used as a feedback device in a servo control type of amplifier/speaker combination.

Anechoic. Anechoic is the term used to describe a reflection-free acoustic space. This is usually thought of as the ideal measurement situation for the design and evaluation of loudspeakers. An anechoic environment can be achieved by having the speaker suspended at least 50 feet from the nearest surface; by constructing a special room or "chamber" which incorporates large amounts of sound dampening material; or by using acoustic analyzers (such as the one used in the preparation of this book), which can "window" or "gate" out reflections.

Asymmetrical. An asymmetrical filter, as applied to crossover networks, refers to a crossover having different slopes on the high-pass section of the network rather than on the low-pass section. An example would be to have a 12dB/octave Butterworth slope on the low-pass section of a crossover, and a 24dB/octave L-R slope on the high-pass section.

Backplate/Frontplate. The backplate and frontplate of a loudspeaker motor are the flat metal discs used to complete the magnetic circuit. They are also used to direct the magnetic field to a focal point in the gap formed by the frontplate and polepiece.

Bandpass Network. Loudspeaker crossovers are frequency-dependent attenuation devices made up of a combination of high-pass, low-pass, and bandpass network circuits. Bandpass networks pass a central "band" or frequency range unaltered and then progressively attenuate higher and lower frequencies.

Complex Impedance. A speaker's impedance is a combination of two elements, DC resistance and AC reactance. Using vector analysis, the simple impedance obtained by the constant voltage or constant current impedance measurement techniques can be broken down into its constituent elements of DC resistance and capacitive and inductive reactance. Mathematically, this amounts to looking at both the real and imaginary vectors.

Cone Excursion. Cone excursion is the term used to describe the piston motion a woofer, midrange, or tweeter exhibits reacting to an input signal. As long as the voice coil remains in the magnetic gap, the cone excursion is regarded as being linear. If the motion allows the voice coil to begin moving any of its total winding outside the gap, the excursion is regarded as nonlinear and is a source of severe distortion.

Cone Velocity. Cone velocity is the rate of cone excursion expressed in meters per second.

Conjugate Networks. Conjugate networks are crossover circuit elements which provide a "mirror" function for the electrical aspect of a speaker's operation. For example, if a speaker impedance has a resonance peak, a circuit LCR type of conjugate would provide a mirror-imaged impedance dip, which in this case would result in the overall impedance being changed from a resonance peak to a flat nominal level.

CR Conjugate. A CR conjugate describes a conjugate circuit containing two elements, a capacitor and a resistor, in series with each other and placed in parallel with the speaker voice coil. Referred to as a "Zobel" network, this type of conjugate is typically used to "mirror" the reactive rise in a speaker's impedance curve, rendering the impedance a flat, nominal value above the frequency where the rise begins.

Cumulative Spectral Decay Curve. A CSD plot is sometimes referred to as a "waterfall" plot, this is a type of "time domain" frequency response decay curve produced by several of the FFT audio analyzers such as the Techron TEF 20 and the DRA Labs' MLSSA. The plot shows the progressive magnitude decay of a transducer stimulated by some type of impulse signal. The plot usually displays the maximum output caused from the stimulus and also the magnitude as it changes (decreases in magnitude, or "decays") over some time period following when the impulse is terminated. The plot sometimes makes it easier to track resonances within a system.

Extensional Damping. One of the strongest "colorations" found in loudspeakers is caused by the cabinet rebroadcasting the information produced by the speaker. This can be minimized by a number of techniques, such as using a stiffer, less resonant box construction material, bracing, or extensional damping. Extensional damping material is placed on the interior surfaces of a loudspeaker cabinet in order to reduce cabinet resonances. Typical materials are rubber pads, lead sheeting, or felt paper.

Free-Air Resonance. Free-air resonance refers to the resonance of a driver (woofer, midrange, or tweeter) when suspended in a frame with no baffle or proximity of large reflective objects.

Groundplane. Groundplane refers to a type of loudspeaker measurement technique. To perform a "groundplane" frequency response measurement, the speaker and microphone are placed on the ground in a large unobstructed area having a smooth, flat, hard surface, such as a parking lot. The measurement is much the same as an anechoic measurement with the addition of the ground "reflection." Because of this reflection, the measurement is 6dB greater in overall magnitude than a standard anechoic measurement taken at the same voltage level and the same microphone distance.

Group Delay. Group delay is the slope of the phase response and is mathematically derived by taking the negative derivative of the phase slope. A flat group delay curve indicates that all frequencies are arriving simultaneously, while a peak in a group delay curve indicates some frequencies arriving later than others.

High-Pass Network. Crossover networks are a type of frequency-dependent attenuation devices made up of a combination of high-pass, low-pass, and bandpass network circuits. High-pass networks progressively attenuate lower frequencies while passing higher frequencies unaltered.

Iterate. Iteration is a mathematical/computer term meaning to "repeat." When you have an optimizing process set up in a loudspeaker computer program to test different values of capacitors, resistors, and inductors in a crossover network, the program will change the values, calculate the outcome, compare it to the desired target, and if necessary, repeat the operation and move on to the next value change (iteration).

LCR Conjugate. A LCR conjugate circuit is a crossover circuit element formed by an inductor, capacitor, and resistor, either in parallel or in series. This circuit configuration can be used to cancel resonance peaks in impedance or to alter peaks in a speaker's frequency response.

Lobing. When two noncoincident drivers (two drivers placed side-by-side, one above the other, and not coaxial) combine within the same frequency range, the phase relationship between the two drivers may cause the radiation pattern exhibited by the combination to tilt, rather than be directed immediately forward of the driver as is the case with a single radiating speaker.

Low-Pass Network. Crossover networks are frequency-dependent attenuation devices made up of a combination of high-pass, low-pass, and bandpass network circuits. Low-pass networks progressively attenuate higher frequencies while passing lower frequencies unaltered.

Noncoincident. This term refers to two drivers mounted side-by-side (one above the other) and thus radiating from two separate and distinct points in space.

Null. A "null" is a condition which exists when two independent signals are typically "out-of-phase" and cancel each other. When the woofer and tweeter in a two-way loudspeaker are correctly "in-phase," reversing the leads on the tweeter (reversing its polarity) will cause a "null" or hole in the frequency response centered near the crossover frequency.

On-Axis/Off-Axis. A line drawn perpendicular to the mounting baffle from the center of a speaker is considered the on-axis position in terms of both listening and measurement. Off-axis is any other position.

Optimization. For the purpose of this book, optimization is the computer process that automatically iterates the values of a crossover network to produce a specified voltage response. The specified voltage response is one which causes the driver's frequency response to achieve a-desired target response.

Q. "Q" is a composite term used to describe resonant magnification. In loudspeaker design, this term generally refers to either the "Q" of a speaker's resonance in free-air, the "Q" of the resonance of a speaker/box combination, or the "Q" of the resonance produced by an LC (inductor/capacitor) circuit used in a crossover network section.

Q_{TC}. Q_{TC} is the total combined "Q" of a speaker when placed in a sealed, closed-box type of enclosure. Q_{TC} is used to define sealed boxes only and is not an appropriate term when discussing the damping characteristics of a vented type of enclosure.

Q_{TS}. Q_{TS} is the total free-air "Q" of a driver and combines both the mechanical "Q" and electrical "Q."

Ripple. This is a term used to define a frequency response which has specific deviations from a perfect, flat magnitude. For example, some vented box alignments produce a flat response except for perhaps some 0.5–1dB deviations above the speaker's roll-off frequency. These deviations are called "ripple."

Transfer Function. A transfer function is the mathematical process which describes the alteration a network performs on an electrical system. Each high-pass, low-pass, and bandpass network used in a loudspeaker crossover can be described as some type of transfer function.

X_{MAX}. X_{MAX} is the measure of the one-direction (from center of the gap) amount of piston movement a speaker can produce with the coil winding still remaining in the magnetic gap. The value of X_{MAX} associated with a given loudspeaker voice coil is calculated by subtracting the total length of the voice coil winding (usually expressed in mm) from the total height of the magnetic gap (also usually expressed in mm), and dividing that number by two.

ADVERTISER FAST FAX

A & S Speakers . . . . (415) 648-5306
LinearX Systems, Inc. . . . . (503) 598-9258
M & C Speakerworks . . . . (619) 536-1128
Madisound . . . . (608) 831-3771
Meniscus . . . . (616) 534-7676
MIT Components . . . . (916) 823-0180
North Creek Music Systems . . . . (315) 369-2500
Parts Express . . . . (513) 222-4644
Solen, Inc. . . . . (514) 443-4949
Speaker City, USA . . . . (818) 846-1009

Rhythm

Our flagship, the **Rhythm** has been precisely engineered from our finest drivers, crossover components, and cabinetry to be one of the most accurate, dynamic, and musically involving loudspeakers available today.

The Drivers: The **Rhythm** features the two most advanced drivers hand made in Denmark by **SCAN-SPEAK**, one of the acknowledged leaders in dynamic driver technology:

The tweeter is the venerable **SCAN-SPEAK D2905** 1" fabric dome, a tweeter widely used in many of Europe's finest systems and considered by many designers, reviewers and audiophiles alike as the most musical tweeter made. The D2905 excels in naturalness, transparency, and soundstaging, and with its superbly designed motor structure surpasses every other tweeter on the market in dynamic capability.

The woofers are the **SCAN-SPEAK 18W/8544** 7" Kevlar® drivers in a symmetric MTM array. These woofers have incredible speed and accuracy, and because of their hand-coated Kevlar® cones are absolutely free of cone break-up and its associated distortions. These drivers reproduce a level of detail that one can usually find only in the finest planer systems.

Also, because of their patented "Symmetric Drive" motor structure, the **SCAN-SPEAK** woofers are far more dynamic than conventional drivers, yet with harmonic and intermodulation distortion that is *an order of magnitude lower* than those from other manufacturers.

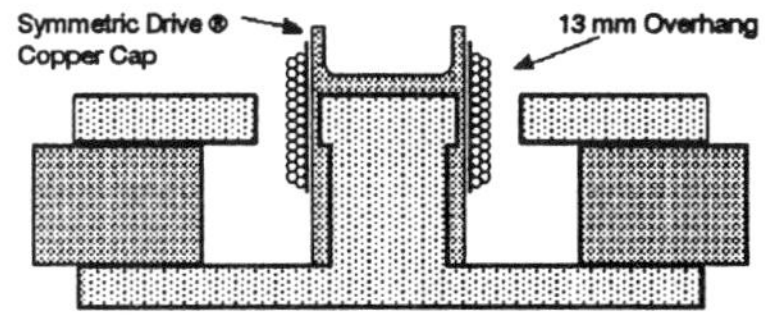

SCAN-SPEAK Motor Structure

The Crossover: The **Rhythm** crossover is a no-compromise symmetric third order network executed with the finest passive crossover components available and provided in ±1% hand matched pairs:

The inductors are **10 AWG North Creek Music Coils**, a hex laid coil featuring a military grade high purity copper conductor and a winding pattern that yields the lowest DC Resistance of any inductor on the market. The 10 AWG Music Coil has been optimized for midrange purity and an unequalled bass line.

The capacitors are **Sprague-Dearborn High Frequency Metallized Polypropylene Film** bypassed by Sprague-Dearborn ***Ultra*-CAP**s. These capacitors have been optimized to operate in 100 kHz switched mode power supplies, and easily surpass even the finest audio-grade polypropylene capacitors in terms of smoothness, detail and transparency. The Sprague capacitors have the most natural midrange and top end of any capacitor we know of.

The resistors are **Ohmite 1% Wirewound Precision Power Resistors.** These resistors are purely resistive to 0.5 MHz, and are dynamically stable to a *full 1000% overload*. The Ohmite resistors are the most transparent of all that we have tested.

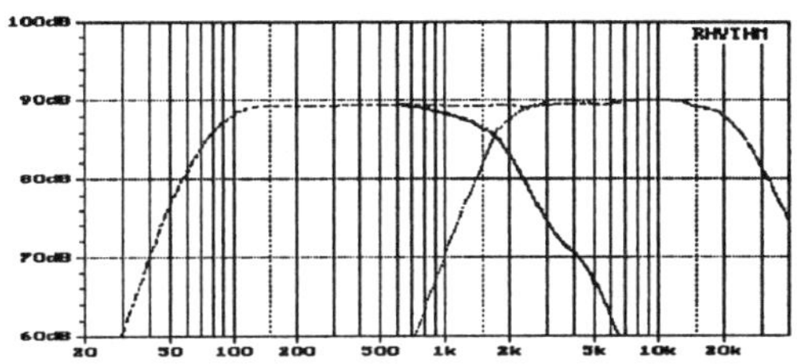

Rhythm Frequency Response

Rhythm internal wiring is manufactured in our own facilities from a combination of **Tef-Flex AG** Silver-plated Copper and ***Vampire* *OFHC*** Copper. The wire gauges and combinations have been optimized for their particular roles within the system.

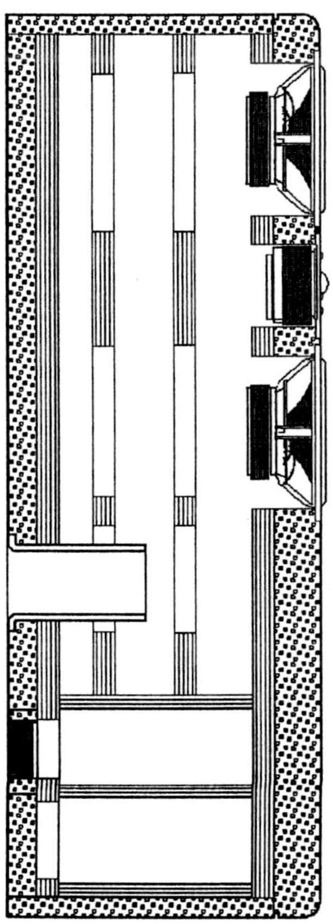

The Cabinet: The **Rhythm** cabinet exemplifies our extraordinary attention to detail. The cabinet features a 2¼" thick front panel and 1⅞" thick rear panel made of a composite of MDF and Baltic Birch plywood. The plywood is imported from Finland and is a lamination of 13 layers of void-free Birch. Bonded to a layer of MDF with NCMS Soft Glue, the resulting panel is incredibly rigid, relatively light and very well damped. Baltic Birch is also used for the bracing structure, and the shell is MDF. The cabinet interior is coated with NCMS "Glop" for panel resonance absorption, and the cabinet base is filled with sand and spiked to the floor. Lastly, the cabinet is finished in natural hardwood veneer.

Owing to its construction, the **Rhythm** cabinet is extremely rigid and incredibly silent. Harmonic colorations due to resonances of cabinet walls in lesser loudspeakers are simply *absent*. Deep and mid-bass is tightly controlled and exceptionally well defined, while the purity of the entire audio spectrum is utterly convincing.

Tonal Accuracy: The **Rhythm** reproduces music with exceptional naturalness. Due to its smooth, extended frequency response and controlled radiation pattern, fundamental and harmonic overtones are reproduced flawlessly throughout the audio spectrum.

Soundstage: The **Rhythm** recreates an enormous soundstage that extends behind, beyond and well above the loudspeaker positions. Specific images within the stage are well focussed, correctly sized, a carry a natural solidity that is completely convincing.

Dynamics: The **Rhythm** is more dynamic than most loudspeakers twice its size. The twin 7" Kevlar cones can move more air than most 10" woofers, but because of their incredibly powerful motor structures they do so in a fraction of the time. The result is a low end with both impact and speed, and an uncolored midrange rich in harmonic detail. The Rhythm has the rare ability to capture and communicate the *essence* of the musical event.

Overview:

Description: Two-way, three driver floor standing loudspeaker system. QB3 Vented, 24l cabinet, Fb = 45 Hz, F3 = 57 Hz.

Driver Complement:
Scan-Speak D2905 1" Fabric Dome Tweeter,
(2) Scan-Speak 18W/8544 7" Kevlar® woofers.

Frequency Response: 57 Hz - 20 kHz, ± 3 dB.

Sensitivity: 89.5dB/2.83V/1m in 4π.

Loudspeaker Dimensions: 9½x12x42.

Kevlar® is a trademark of DuPont

Finishes:

The standard finish is a rift-sawn White Oak with gold flakes. Optional veneers and custom finishes are available upon request.

For a complementary catalog, please call or write:

North Creek Music Systems
Main Street, PO Box 1120-R
Old Forge, NY 13420
Voice/Fax (315) 369-2500

Simply Better Technology

NEW 1994 CATALOG
NEW 1994 CATALOG

All a Speaker Builder Needs to Know About Dual Voice Coil Woofers

Model	81524DVC	10204DVC	10207DVC	1052DVC	12204DVC	1252DVC	15254DVC	15258DVC
Size Inches	8	10	10	10	12	12	15	15
Fs Hz	31.7	21.2	18.6	20.4	22.8	15	23	22.5
Mmd g	38	50.4	57	46	68.8	78	122	121.5
Cms (μm/N)	631.44	1045.4	1138.6	1220.1	550.6	1331.4	346.1	367.38
Vas Liters	39.2	168	184	197	220	533	347	368
Rscc Ω per coil	3.5	3.6	5.7	6.1	3.6	5.6	3.7	5.5
VcL Mh @1Khz	.34	.35	.51	.46	.26	.3	.25	.36
Qms	9.2	3.5	3.43	3.68	4.58	4.1	5.71	5.35
Qes	.32	.21	.23	.28	.42	.39	.47	.52
Qts	.31	.2	.22	.26	.38	.36	.44	.47
Xmax mm peak	5	5	5	6	5	6	5.5	5.5
Efficiency dB, both coils driven (2.83V/M)	92	93.5	92	92	94.5	92	96	93.5
Response	30-2.3K	30-1.5K	30-1.5K	25-2.5K	25-1.5K	20-1.8K	30-1.5K	30-1.5K
Power Watts (coil 1 / coil 2)	80 (40/40)	200 (100/100)	200 (100/100)	100 (50/50)	200 (100/100)	100 (50/50)	200 (100/100)	200 (100/100)
Magnet Wt. oz.	20	40	40	30	40	30	60	60
All units have Black Polypropylene Cones, foam surrounds, and voice coils made of high temperature aluminum on Kapton formers.								
Voice Coil	1.5"	2"	2"	1.5"	2"	1.5"	2"	2"
Cutout & Depth	7.12" 3.37"	9.12" 4.45"	9.12" 4.45"	9.12" 4.45"	11.11" 5"	11.11" 5"	13.87" 6"	13.87" 6"
Price Each	**$34**	**$49**	**$49**	**$41**	**$53**	**$42**	**$73**	**$73**

Alignments	81524DVC			10204DVC			10207DVC			1052DVC			12204DVC					1252DVC				15254DVC				15258DVC			
Rg (Ind. DCR)	0	.5	1	0	.5	1	0	.5	1	0	.5	1	0	.5	1	0	.5	0	0	0	0	0	.5	0	.5	0	.5	0	.5
Vb Liters	22	33	46	21	30	42	26	33	40	47	57	69	85	85	100	113	142	85	100	130	142	100	100	142	142	100	100	142	142
F3 Hz	39	34	30	52	45	39	45	41	37	38	35	33	42	38	37	31	28	32	31	30	26	40	37	38	35	38	37	35	34
Fb Hz	34	33	30	41	37	33	35	32	30	31	29	27	Sealed			24	21	Sealed			17	Sealed				Sealed			
Port Dia. In.	2	2	2	2.5	2.5	2.5	2.5	2.5	2.5	3	3	3	Qtc .75	Qtc .85	Qtc .8	3	3	Qtc .96	Qtc .9	Qtc .8	3	Qtc .92	Qtc 1	Qtc .8	Qtc .91	Qtc 1	Qtc 1.1	Qtc .9	Qtc 1
Port Length In.	6	5	4.3	8.7	7.4	6.3	9.5	9	8.6	9.9	9.3	8.6	Sealed			6.1	5.9	Sealed			11	Sealed				Sealed			

4th Order Bandpass	81524DVC			10204DVC			10207DVC			1052DVC			12204DVC			1252DVC			15254DVC			15258DVC		
Alignment Type *	BES	BUT	CHV	BES	BUT	CHV	BES	BUT	CHV	BES	BUT	CHV	BES	BUT	CHV	BES	BUT	CHV	BES	BUT	CHV	BES	BUT	CHV
Vb Rear Sealed ft^3	.7	.4	.17	.8	.5	.25	1	.68	.32	1.75	1	.5	6.4	3.4	1.4	x	x	2.75	x	7.5	2.9	x	10.1	3.65
Vb Front Vented ft^3	.46	.3	.15	.7	.47	.25	.9	.61	.31	1.4	.95	.47	3.7	2.4	1.2	x	x	2.4	x	4.6	2.3	x	5.7	2.8
Fb (Hz)	55	67.3	95	62	76	107	51	62	88	45	56	79	35	43	60	x	x	42	x	37.5	53	x	34	48
Vent Diameter In.	3	2.5	2	4	3	3	4	3	3	4	4	3	2x4	2x4	2x3	x	x	2x3	x	2x4	2x4	x	2x4	2x4
Vent Length In.	10.8	7.4	4.4	8.9	4.3	4	10.8	5.4	5.2	8.3	7.75	4.2	11.2	11.5	6	x	x	6.2	x	6.75	6.75	x	6.6	6.8
F3 Lower	36	40	50	40	45	58	33	38	48	30	27	44	23	27	33	x	x	23	x	23	28	x	22	27
F3 Higher	80	105	155	90	110	180	75	100	155	65	68	125	50	68	103	x	x	75	x	60	95	x	55	85

We thought that Speaker Builders would appreciate a chart of Madisound Dual Voice Coil Woofers with specifications and applications. There are several reasons why you may wish to choose a woofer with two voice coils:

1) You want to use a common subwoofer and drive it with a stereo amplifier.
2) You wish to parallel the voice coils at low frequencies to give added bass boost to your system.
3) You may use an impedance not commonly available. Two 4 ohm coils can be paralleled for a 2 ohm net impedance. Two 8 ohm coils can be used in series for a 16 ohm load.
4) Many Madisound dual voice coil woofers can be used in very small boxes. This is ideal for quality autosound bass and unobtrusive video systems.
5) Adding a subwoofer is not expensive; low frequency crossover filters are available starting at $50 a pair, (even less if you buy parts for home assembly). Cabinets are limited only to your imagination.
6) A subwoofer is the most dramatic improvement you can make to the average existing stereo system.

* BES: Bessel; BUT: Butterworth; CHV: Chebychev

Ordering Information: All speaker orders will be shipped promptly, if possible by UPS. COD requires a 25% prepayment, and personal checks must clear before shipment. Add 10% for shipping, residents of Alaska, Canada and Hawaii, and those who require Blue Label air service, please add 25%. There is no fee for packaging or handling, and we will refund to the exact shipping charge. We accept Mastercard or Visa on mail or phone orders.

Madisound Speaker Components
(8608 University Green)
P.O. Box 44283
Madison, WI 53744-4283 U.S.A
Voice: 608-831-3433
Fax: 608-831-3771

MW 114-S

Neodymium Magnet DPC Cone 4" Woofer

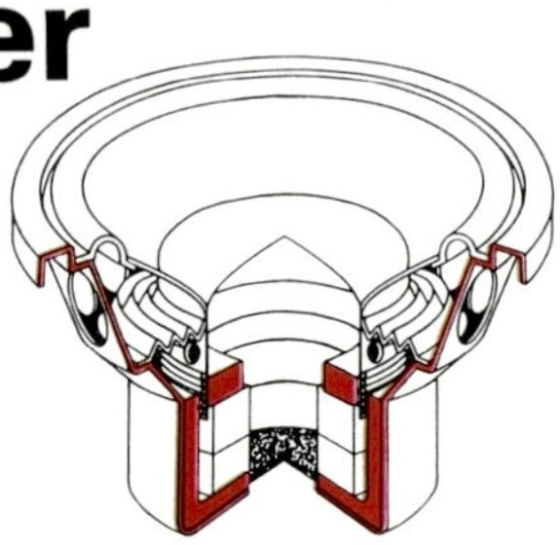

Specification

Overall Dimensions	**Ø118mm (4.64") x 58mm(2.29")**
Mounting Baffle Hole Diameter	**Ø95mm (3.75")**
Magnet System	**Pot Type, Vented, Neodymium Magnet**
Nominal Power Handling (Din)	**150W**
Transient Power - 10ms	**800W**
Voice Coil Diameter	**54mm (2.125")**
Voice Coil Type/Former	**Hexatech Aluminium**
Frequency Response	**55-7000 Hz**
FS - Resonant Frequency	**65 Hz**
Sensitivity 1W/1m	**87 dB**
Z - Nominal Impedance	**8 ohms**
RE - DC Resistance	**5.6 ohms**
LBM - Voice Coil Inductance @ 1kHz	**0.47 mH**
Magnetic Gap Width	**1.25mm (0.050")**
HE - Magnetic Gap Height	**6mm (0.236")**
Voice Coil Height	**12mm (0.472")**
X - Max. Linear Excursion	**3mm**
B - Flux Density	**0.88T**
BL Product (BXL)	**6.75**
Qms - Mechanical Q Factor	**2.32**
Qes - Electrical Q Factor	**0.36**
Q/T - Total Q Factor	**0.31**
Vas - Equivalent Cas Air Load	**3.18 litres (0.113 cu. ft.)**
MMS - Moving Mass	**7.00gm**
CMS	**807μm/n**
SD - Effective Cone/Dome Area	**53cm² (20.86 sq. in.)**
Cone/Dome Material	**DPC (Damped Polymer Composite)**
Nett Weight	**0.500 kg**

Specifications given are as after at least 45 minutes of high power, low frequency running, or 24 hours normal power operation.

The 114-S is the first of Morel's new generation of woofers, featuring a powerful Neodymium magnet system which provides increased sensitivity, lower Qt and reduced distortion. For a 4" driver it is unique in having a large 54mm (2.125") diameter Hexatech aluminium voice coil.

Benefits of this large voice coil diameter include a very high power handling capacity and lack of sound level compression. In addition, it allows the use of a very shallow cone profile. Coupled with the use of Damped Polymer Composite cone material and a rubber surround, this provides excellent dispersion (off-axis response), resistance to cone break-up (even at high sound pressure levels) and lack of colouration.

Frequency and phase response are very flat, while the roll-off is very smooth. The MW 114-S may be used either as a bass-mid range in 2-way systems, or as a mid-range in multi-way systems.

The vented magnet system is encased within a steel chassis, which improves efficiency and shields the magnet, virtually eliminating stray magnetic fields. The MW 114-S is ideal not only for high quality hi-fi, but also TV, video and surround-sound applications.

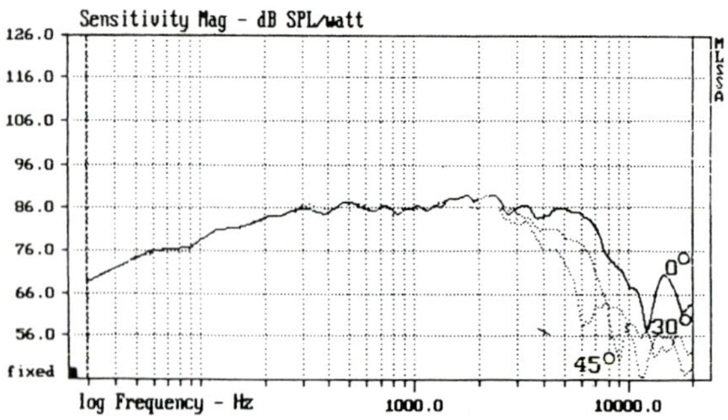

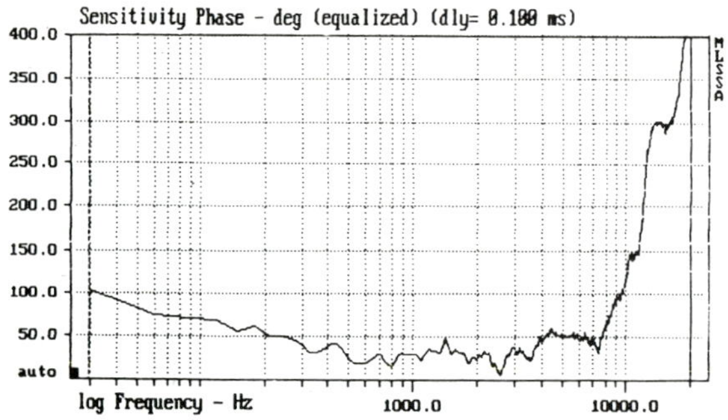

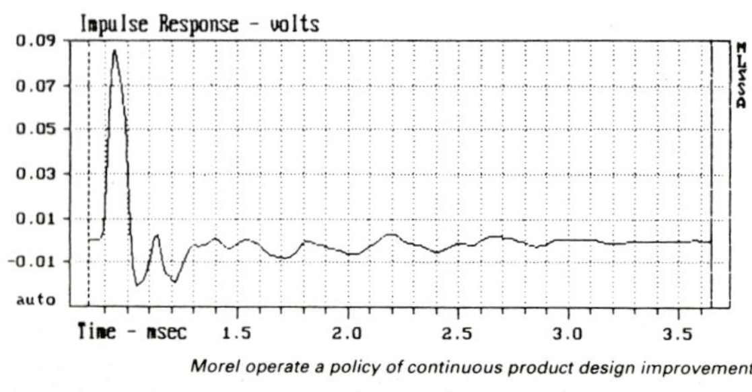

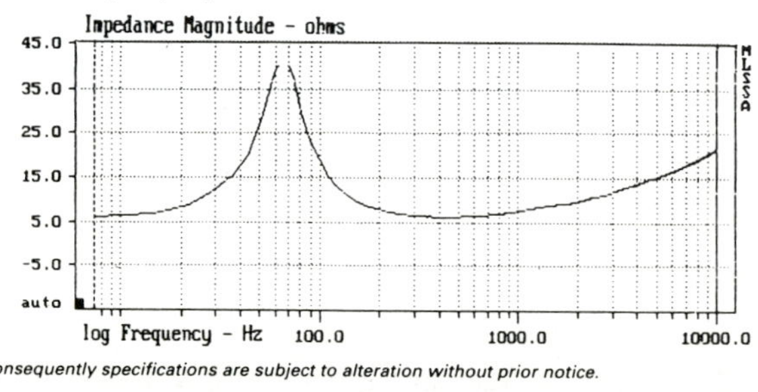

Morel operate a policy of continuous product design improvement, consequently specifications are subject to alteration without prior notice.

high fidelit
range

morel (U.K.) ltd.
11 Foxtail Road,
Nacton Rd. (Ransomes)
Industrial Estate,
Ipswich, IP3 9RT
England
Tel: (0473) 719212
FAX: (0473) 716206

morel acoustics USA
414 Harvard Street
Brookline, MA 02146
Tel: (617) 277-6663
FAX: (617) 277-2415

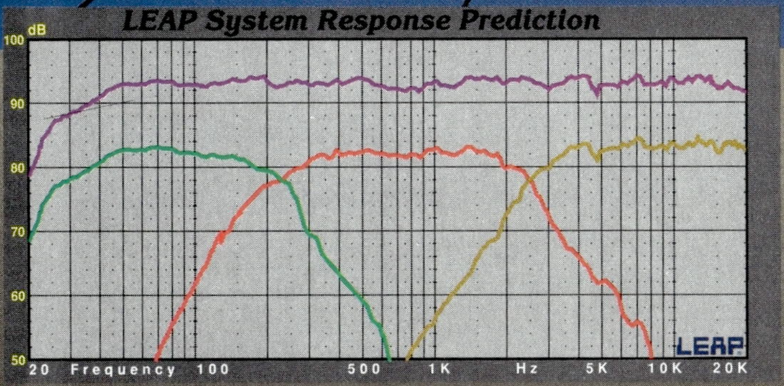

LEAP System Response Prediction
dB
100
90
80
70
60
50
20 Frequency 100 500 1K Hz 5K 10K 20K
LEAP

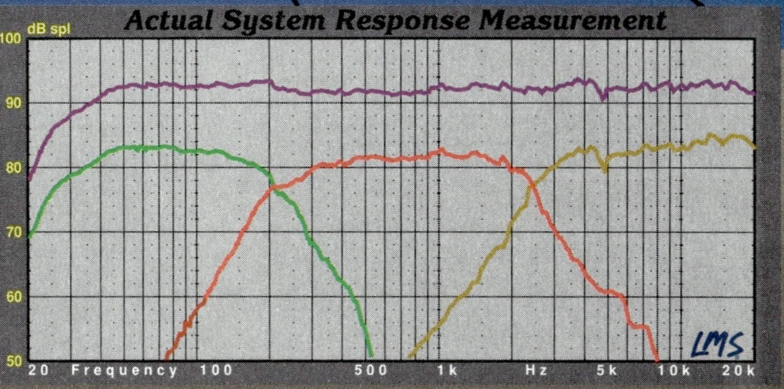

Actual System Response Measurement
dB spl
100
90
80
70
60
50
20 Frequency 100 500 1k Hz 5k 10k 20k
LMS